Student Solutions Manual

Michael Kenney
Michigan State University

Paul Gaus
College of Wooster

CHEMISTRY
The Study of Matter and Its Changes

Second Edition

James E. Brady
St. John's University, New York

John R. Holum
Augsburg College (Emeritus), Minnesota

John Wiley & Sons, Inc.
New York / Chichester / Brisbane / Toronto / Singapore

ISBN 0-471-12076-6

Printed in the United States of America

10 9 8 7 6 5 4 3 2 1

Preface

Congratulations! By investing in this Student Solutions Manual you have demonstrated a desire to go beyond the normal effort expended in order to understand the material presented in your class and the textbook. This solutions manual has been written to assist you as you explore the world around you from the perspective of a chemist.

Chemistry, for many people, is a strange and foreign language that will take time to comprehend and appreciate. It is filled with new vocabulary and ideas. It appears, to many, to be filled with abstract ideas and volumes of equations and symbols that are impossible to understand, let alone, commit to memory. Yet, an understanding of chemical phenomena and ideas will help you to understand the way the natural world works. A study of chemistry will provide you with a logical method of approaching a variety of problems. It will give you tools that may help you in a variety of tasks you will encounter throughout your lifetime.

In order to succeed in your studies it is imperative that you continually think about the material you are discussing in class, reading in the text or using to solve a problem. The information provided is intended to guide you to a specific point but it is up to you to follow the appropriate paths in order to find a reasonable solution.

This solutions manual provides completely worked out solutions to many of the exercises found at the end of each chapter of the text. I have tried to work each problem in a clear and logical fashion. Often, I have included commentary to help you to "see" the logic of a particular solution or to explain "why" a certain tact in solving a problem was used. I have tried to be consistent in the method used to solve problems and I have used dimensional analysis whenever appropriate.

The purpose of this manual is to help you to understand the concepts you are discussing and reading. It is intended to be used as a tool as you endeavor to understand the material. However, there are good and bad ways to use the information presented within these pages. When approaching a problem for the first time I recommend the following procedure:

- read the problem in order to determine the specific question being asked.
- restate the question in your own words.
- attempt to solve the problem using the tools you possess and learn in the course.
- check your solution to insure that it is reasonable, i.e., is it of the correct magnitude?, are the units correct?, is your solution the answer to the question that was asked?
- if you are unsure of the correctness of the answer or the units do not "work out" or you answered a different question than the one asked, seek assistance from a classmate, an instructor or this manual.
- always rework any problems that were difficult to understand.

Finally, I want to wish you good luck. I hope that you enjoy your chemistry course. I hope that you will find enjoyment as you embark upon a journey that will, hopefully, provide you with an understanding of the world around you.

Acknowledgments

I first want to thank my wife Cyndi. Her support and encouragement is the reason this project is completed. I also want to thank my editor Joan Kalkut. She has helped me immensely throughout this project. Bill Draper and Laura Swart-Hills provided much needed assistance when it was really needed. I especially want to thank the best proof readers I have ever had the pleasure to work with: Susan Holladay, Nicole Carlson and Kirsten Lowrey. They worked each and every problem in order to insure accuracy and completeness. They checked my work and caught my mistakes. They were generous in their suggestions of alternative wordings. They persevered even when deadlines approached and packages did not. Most importantly, they reminded me to always remember the students. Without their help, this manual would not be.

Michael J. Kenney
October 1995

Table of Contents

CHAPTER ONE
Review Exercises

1.16   (a)   0.01 10^{-2}

         (b)   0.001 10^{-3}

         (c)   1000 10^{3}

         (d)   0.000001 10^{-6}

         (e)   0.000000001 10^{-9}

         (f)   0.000000000001 10^{-12}

         (g)   1000000 10^{6}

1.18   (a)   0.01   (d)   0.1

         (b)   1000   (e)   0.001

         (c) 10^{12} (f)   ~~0.001~~ 0.01

1.24   For these problems, use equations 1.1 and 1.2

         (a)   °F = 9/5(°C) + 32 = 9/5(60) + 32 = 140 °F, when rounded to the proper number (two) of significant figures.

         (b)   °F = 9/5(°C) + 32 = 9/5(20) + 32 = 68 °F

         (c)   °C = 5/9(°F − 32) = 5/9(45.5 − 32) = 7.5 °C

         (d)   °C = 5/9(°F − 32) = 5/9(59 − 32) = 15 °C

         (e)   K = °C + 273 = 40 + 273 = 313 K

         (f)   K = °C + 273 = −20 + 273 = 253 K

1.26   °F = 9/5(°C) + 32 = 9/5(37.46) + 32 = 99.43 °F

1.28   °C = K − 273 = 4 − 273 = −269 °C
        °F = 9/5(°C) + 32 = 9/5(−269) + 32 = −452 °F

1.30   K = °C + 273 = −183 + 273 = 90 K

1.33   (a)   four   (b)   five   (c)   four

         (d)   two   (e)   four   (f)   one

1.35   (a)   0.69   (b)   83.24   (c)   0.006

       (d)   22.84   (e)   775.4   (Note: The subtraction gives 107.3 so 775.1 is also a possible answer.)

1.37   (a) 2.34×10^{3} (b) 3.10×10^{7} (c) 2.87×10^{-4}

       (d) 4.50×10^{4} (e) 4.00×10^{-6} (f) 3.24×10^{5}

1.39   (a)   210,000   (b)   0.00000335   (c)   3800

       (d)   0.0000000000046   (e)   0.00000346   (f)   850,000,000

1.41   (a) 2.0×10^{4}

       (b) 8.0×10^{7}

       (c) 1.0×10^{3}

       (d) 2.4×10^{5}

       (e) 2.0×10^{18}

1.46   (a) $\# \text{ km} = (32.0 \text{ dm})\left(\dfrac{1 \text{ m}}{10 \text{ dm}}\right)\left(\dfrac{1 \text{ km}}{1000 \text{ m}}\right) = 3.20 \times 10^{-3} \text{ km}$

       (b) $\# \mu\text{g} = (8.2 \text{ mg})\left(\dfrac{1 \text{ g}}{1000 \text{ mg}}\right)\left(\dfrac{1 \times 10^{6} \ \mu\text{g}}{1 \text{ g}}\right) = 8.2 \times 10^{3} \ \mu\text{g}$

       (c) $\# \text{ kg} = (75.3 \text{ mg})\left(\dfrac{1 \text{ g}}{1000 \text{ mg}}\right)\left(\dfrac{1 \text{ kg}}{1000 \text{ g}}\right) = 7.53 \times 10^{-5} \text{ kg}$

       (d) $\# \text{ L} = (137.5 \text{ mL})\left(\dfrac{1 \text{ L}}{1000 \text{ mL}}\right) = 0.1375 \text{ L}$

       (e) $\# \text{ mL} = (0.025 \text{ L})\left(\dfrac{1000 \text{ mL}}{1 \text{ L}}\right) = 25 \text{ mL}$

       (f) $\# \text{ dm} = (342 \text{ pm})\left(\dfrac{1 \text{ m}}{1 \times 10^{12} \text{ pm}}\right)\left(\dfrac{10 \text{ dm}}{1 \text{ m}}\right) = 3.42 \times 10^{-9} \text{ dm}$

1.48   (a) $\# \text{ cm } = (36 \text{ in.})\left(\dfrac{2.54 \text{ cm}}{1 \text{ in.}}\right) = 91 \text{ cm}$

       (b) $\# \text{ kg } = (5.0 \text{ lb})\left(\dfrac{1 \text{ kg}}{2.205 \text{ lb}}\right) = 2.3 \text{ kg}$

       (c) $\# \text{ mL } = (3.0 \text{ qt})\left(\dfrac{946.4 \text{ mL}}{1 \text{ qt}}\right) = 2800 \text{ mL}$

       (d) $\# \text{ mL } = (8 \text{ oz})\left(\dfrac{29.6 \text{ mL}}{1 \text{ oz}}\right) = 200 \text{ mL}$

       (e) $\# \text{ km / hr } = (55 \text{ mi / hr})\left(\dfrac{1.609 \text{ km}}{1 \text{ mi}}\right) = 88 \text{ km / hr}$

       (f) $\# \text{ km } = (50.0 \text{ mi})\left(\dfrac{1.609 \text{ km}}{1 \text{ mi}}\right) = 80.5 \text{ km}$

1.50 $\# \text{ mL } = (12 \text{ oz})\left(\dfrac{29.6 \text{ mL}}{1 \text{ oz}}\right) = 360 \text{ mL}$

1.52 $\# \text{ lb } = (1000 \text{ kg})\left(\dfrac{2.205 \text{ lb}}{1 \text{ kg}}\right) = 2205 \text{ lb}$

1.54   This individual is 74 in. tall, (6 ft)(12 in./ft) + 2 in.

       $\# \text{ cm } = (74 \text{ in.})\left(\dfrac{2.54 \text{ cm}}{1 \text{ in.}}\right) = 190 \text{ cm}$

1.56   (a) $\# \text{ m}^2 = (6.2 \text{ yd}^2)\left(\dfrac{0.9144 \text{ m}}{1 \text{ yd}}\right)^2 = 5.2 \text{ m}^2$

       (b) $\# \text{ mm}^2 = (4.8 \text{ in.}^2)\left(\dfrac{1 \text{ m}}{39.37 \text{ in.}}\right)^2\left(\dfrac{1000 \text{ mm}}{1 \text{ m}}\right)^2 = 3100 \text{ mm}^2$

       (c) $\# \text{ L } = (3.7 \text{ ft}^3)\left(\dfrac{12 \text{ in.}}{1 \text{ ft}}\right)^3\left(\dfrac{2.54 \text{ cm}}{1 \text{ in.}}\right)^3\left(\dfrac{1 \text{ mL}}{1 \text{ cm}^3}\right)\left(\dfrac{1 \text{ L}}{1000 \text{ mL}}\right) = 1.0 \times 10^2 \text{ L}$

1.58 $\dfrac{\text{\# km}}{\text{hr}} = \left(\dfrac{2235\ \text{ft}}{\text{s}}\right)\left(\dfrac{12\ \text{in.}}{1\ \text{ft}}\right)\left(\dfrac{2.54\ \text{cm}}{1\ \text{in.}}\right)\left(\dfrac{1\ \text{m}}{100\ \text{cm}}\right)\left(\dfrac{1\ \text{km}}{1000\ \text{m}}\right)\left(\dfrac{3600\ \text{s}}{1\ \text{hr}}\right) = 2452\ \dfrac{\text{km}}{\text{hr}}$

1.60 $\dfrac{\text{\# km}}{\text{hr}} = \left(\dfrac{65\ \text{miles}}{\text{hr}}\right)\left(\dfrac{1.609\ \text{km}}{1\ \text{mile}}\right) = 1.0 \times 10^2\ \dfrac{\text{km}}{\text{hr}}$

1.69 The carbon dioxide and water vapor have less potential energy, because the reaction of oxygen with gasoline (combustion) releases potential energy when the products are formed. The excess potential energy of the gasoline and oxygen is transformed to heat.

1.75 $\text{\# kg} = (18\ \text{gal})\left(\dfrac{3.786\ \text{L}}{1\ \text{gal}}\right)\left(\dfrac{1000\ \text{mL}}{1\ \text{L}}\right)\left(\dfrac{0.65\ \text{g}}{1\ \text{mL}}\right)\left(\dfrac{1\ \text{kg}}{1000\ \text{g}}\right) = 44\ \text{kg}$

44 kg is equivalent to 97 lbs.

1.77 $\text{\# mL} = (25.0\ \text{g}) \div (0.791\ \text{g/mL}) = 31.6\ \text{mL}$

1.79 mass of silver = 62.00 g – 27.35 g = 34.65 g
volume of silver = 18.3 mL – 15.0 mL = 3.3 mL
density of silver = (mass of silver) ÷ (volume of silver)
= 34.65 g ÷ 3.3 mL = 11 g/mL

1.81 $\text{sp. gr.} = \dfrac{d_{\text{substance}}}{d_{\text{water}}} = \dfrac{8.65\ \text{lb}/\text{gal}}{8.34\ \text{lb}/\text{gal}} = 1.04$

1.83 The density of gold is 1.20×10^3 lb/ft^3 using equation 1.10.
$\text{\# lbs} = (1\ \text{ft}^3)(1.20 \times 10^3\ \text{lb/ft}^3) = 1.20 \times 10^3\ \text{lb}.$

1.85 Since the density closely matches the known value, we conclude that this is an authentic sample of ethylene glycol.

1.89 (a) In order to determine the volume of the pycnometer, we need to determine the volume of water that fills it. We will do this using the mass ofg the water and its density.

mass of water = mass of filled pycnometer – mass of empty pycnometer
= 36.842 g – 27.314 g
= 9.528 g

$\text{volume} = (9.528\ \text{g})\left(\dfrac{1\ \text{mL}}{0.99704\ \text{g}}\right) = 9.556\ \text{mL}$

(b) We know the volume of chloroform from part (a). The mass of chloroform is determined in the same manner that we determined the mass of water;

mass of chloroform = mass of filled pycnometer – mass of empty pycnometer
= 41.428 g – 27.314 g
= 14.114 g

Density of chloroform = $(14.114 \text{ g}) \div (9.556 \text{ mL}) = 1.477 \text{ g mL}^{-1}$

1.94 (a) $\dfrac{\# \text{ g}}{\text{L}} = \left(\dfrac{1.040 \text{ g}}{\text{mL}}\right)\left(\dfrac{1000 \text{ mL}}{\text{L}}\right) = 1040 \text{ g}/\text{L} = 1.040 \text{ X } 10^3 \text{ g}/\text{L}$

(b) $\dfrac{\# \text{ kg}}{\text{m}^3} = \left(\dfrac{1.040 \text{ g}}{\text{mL}}\right)\left(\dfrac{1 \text{ kg}}{1000 \text{ g}}\right)\left(\dfrac{1 \text{ mL}}{1 \text{ cm}^3}\right)\left(\dfrac{100 \text{ cm}}{1 \text{ m}}\right)^3$

$= 1040 \text{ kg}/\text{m}^3 = 1.040 \text{ X } 10^3 \text{ kg}/\text{m}^3$

1.96 First we need to recall the formula which describes the circumference of a circle:
circumference $= 4\pi r = 4\pi(12 \text{ in.}) = 1.50 \text{ X } 10^2$ in. The circumference is the total distance a point on the edge travels in each revolution. To calculate speed:

$$\dfrac{\# \text{ miles}}{\text{hr}} = \left(\dfrac{33\frac{1}{3} \text{ revolutions}}{1 \text{ minute}}\right)\left(\dfrac{1.50 \text{ X } 10^2 \text{ in.}}{1 \text{ revolution}}\right)\left(\dfrac{2.54 \text{ cm}}{1 \text{ in.}}\right)\left(\dfrac{1 \text{ m}}{100 \text{ cm}}\right)$$

$$\times \left(\dfrac{1 \text{ km}}{1000 \text{ m}}\right)\left(\dfrac{1 \text{ mile}}{1.609 \text{ km}}\right)\left(\dfrac{60 \text{ min}}{1 \text{ hr}}\right)$$

$$= 4.74 \text{ miles / hr}$$

1.98 (a) The average kinetic energy of the system must decrease since the system becomes "cool," i.e., the temperature decreases.

(b) The total heat of this system decreases as the system cools. Therefore, the total kinetic energy must decrease.

(c) Since energy must be conserved in this system (an insulated container), the potential energy must increase as a result of the decrease in kinetic energy.

1.100 (a) The only temperature at which any absolute scale overlaps any other absolute scale is zero.

(b) We will start with equation 1.1; $t_F = \left(\dfrac{9\ ^\circ F}{5\ ^\circ C}\right)t_C + 32\ ^\circ F$. Now, since one fahrenheit degree unit is equivalent to one Rankine unit and one celcius degree unit is equivalent to one Kelvin unit, we can rewrite equation 1.1 as;

$$t_R = \left(\frac{9\ R}{5\ K}\right)t_K + 32\ R = \left(\frac{9\ R}{5\ K}\right)(373\ K) + 32\ R = 703\ R$$

where we have used the value of 373 K as the boiling point of water on the Kelvin scale.

CHAPTER TWO
Review Exercises

2.8 (a) 2 K, 2 C, 4 O
(b) 2 H, 1 S, 3 O
(c) 12 C, 26 H
(d) 4 H, 2 C, 2 O
(e) 9 H, 2 N, 1 P, 4 O
(f) 4 C, 10 H, 1 O

2.10 (a) 1 Ni, 2 Cl, 8 O
(b) 1 Cu, 1 C, 3 O
(c) 2 K, 2 Cr, 7 O
(d) 2 C, 4 H, 2 O
(e) 2 N, 9 H, 1 P, 4 O
(f) 3 C, 8 H, 3 O

2.15 (a) 6 N, 3 O
(b) 14 C, 28 H, 14 O
(c) 4 Na, 4 H, 4 C, 12 O
(d) 4 N, 8 H, 2 C, 2 O
(e) 2 Cu, 2 S, 18 O, 20 H
(f) 10 K, 10 Cr, 35 O

2.16 (a) 6 (b) 3 (c) 27

2.24 An authentic sample of laughing gas must have a mass ratio of nitrogen/oxygen of 1.75 to 1.00. The only possibility in this list is item (c), which has the ratio of mass of nitrogen to mass of oxygen of 8.84/5.05 = 1.75.

2.26 (a) This ratio should be 4/2 = 2/1, as required by the formulas of the two compounds.
(b) Twice 0.597 g Cl, or 1.19 g Cl.

2.28 $12 \times 1.660566520 \times 10^{-24}$ g C = $1.998079824 \times 10^{-23}$ g C

2.30 Regardless of the definition, the ratio of the mass of hydrogen to that of oxygen would be the same. If C–12 were assigned a mass of 24 (twice its accepted value), then hydrogen would also have a mass twice its current value, or 2.0158 u.

2.32 $(0.7899 \times 23.9850$ u$) + (0.1000 \times 24.9858$ u$) + (0.1101 \times 25.9826$ u$) = 24.31$ u

2.41 (a) $^{131}_{53}I$ (b) $^{90}_{38}Sr$ (c) $^{137}_{55}Ce$ (d) $^{18}_{9}F$

2.42

	neutrons	protons	electrons
(a)	138	88	88
(b)	8	6	6
(c)	124	82	82
(d)	12	11	11

2.44

	neutrons	protons	electrons
(a)	46	35	36
(b)	32	26	23
(c)	34	29	27
(d)	50	37	36

2.51 Strontium and calcium are in the same Group of the periodic table, so they are expected to have similar chemical properties. Strontium should therefore form compounds that are similar to those of calcium, including the sorts of compounds found in bone.

2.53 Cadmium is in the same periodic table Group as zinc, but silver is not. Therefore, cadmium would be expected to have properties similar to those of zinc, whereas silver would not.

2.69 copper and gold

2.78 (a) Cl_2 (b) S_8 (c) P_4
(d) N_2 (e) O_2 (f) H_2

2.80 HAt

2.82 Bi_2O_3 and Bi_2O_5

2.84 $C_{10}H_{22}$

2.94 These ions are Ca^{2+} and Cl^-.

2.97 This requires a gain of three electrons. There are thus 7 protons and 10 electrons.

2.99 Rubidium, Rb, is in group 1 and has a +1 charge. When it combines with Cl, possessing a −1 charge, the formula would be RbCl. In the second compound, the metal atom should be written first; Na_2S.

2.106 (a) NaBr (b) KI (c) BaO (d) $MgBr_2$ (e) BaF_2

Chapter Two

2.111 (a) KNO_3 (b) $Ca(C_2H_3O_2)_2$ (c) NH_4Cl
      (d) $Fe_2(CO_3)_3$ (e) $Mg_3(PO_4)_2$

2.113 (a) PbO and PbO_2 (b) SnO and SnO_2 (c) MnO and Mn_2O_3
      (d) FeO and Fe_2O_3 (e) Cu_2O and CuO

2.115 (a) $Ca(s) + Cl_2(g) \rightarrow CaCl_2(s)$
      (b) $2Mg(s) + O_2(g) \rightarrow 2MgO(s)$
      (c) $4Al(s) + 3O_2(g) \rightarrow 2Al_2O_3(s)$
      (d) $S(s) + 2Na(s) \rightarrow Na_2S(s)$ or $S_8(s) + 16Na(s) \rightarrow 8Na_2S(s)$

2.122 (a) calcium sulfide   (e) sodium phosphide
      (b) sodium fluoride   (f) lithium nitride
      (c) aluminum bromide   (g) barium arsenide
      (d) magnesium carbide   (h) aluminum oxide

2.124 (a) silicon dioxide   (e) tetraphosphorus decaoxide
      (b) chlorine trifluoride   (f) dinitrogen pentaoxide
      (c) xenon tetrafluoride   (g) dichlorine heptaoxide
      (d) disulfur dichloride   (h) arsenic pentachloride

2.126 (a) periodic acid   (d) hypoiodous acid
      (b) iodic acid   (e) hydroiodic acid
      (c) iodous acid

2.128 (a) sodium nitrite   (e) barium sulfate
      (b) potassium phosphate   (f) ferric carbonate; iron(III) carbonate
      (c) potassium permanganate   (g) potassium thiocyanate
      (d) ammonium acetate   (h) sodium thiosulfate

2.130 (a) chromous chloride; chromium(II) chloride
      (b) ammonium acetate   (k) cobaltous acetate; cobalt(II) acetate
      (c) potassium iodate   (l) auric sulfide; gold(III) sulfide
      (d) chlorous acid   (m) aurous sulfide; gold(I) sulfide
      (e) calcium sulfite   (n) germanium tetrabromide
      (f) silver cyanide   (o) potassium chromate
      (g) zinc(II) bromide   (p) ferrous hydroxide; iron(II) hydroxide
      (h) hydrogen selenide   (q) diiodine tetraoxide
      (i) hydroselenic acid   (r) tetraiodine nonaoxide
      (j) vanadium(III) nitrate   (s) tetraphosphorus triselenide

9

2.131 (a) Na_2HPO_4 (e) $Ni(CN)_2$ (i) Al_2Cl_6 (m) NH_4SCN

(b) Li_2Se (f) Fe_2O_3 (j) As_4O_{10} (n) $K_2S_2O_3$

(c) NaH (g) SnS_2 (k) $Mg(OH)_2$

(d) $Cr(C_2H_3O_2)_3$ (h) SbF_5 (l) $Cu(HSO_4)_2$

2.133 (a) $Au(NO_3)_3$ (f) Sb_2S_3 (k) $HgCl_2$ (p) $(NH_4)_2Cr_2O_7$

(b) $CuSO_4$ (g) $PtCl_2$ (l) $Sr_3(PO_4)_2$ (q) I_2O_5

(c) NH_4BrO_3 (h) CdS (m) Ba_3As_2 (r) P_4S_7

(d) $Ni(IO_3)_2$ (i) $Ba(C_2H_3O_2)_2$ (n) $Co(OH)_2$ (s) S_2F_{10}

(e) PbO_2 (j) Hg_2Cl_2 (o) $Al_2(SO_3)_3$

2.135 (a) sodium bicarbonate

(b) potassium dihydrogen phosphate

(c) ammonium hydrogen phosphate

2.137 (a) First calculate the number of moles of oxygen that are combined with X:

$$(1.14 \text{ g oxygen})\left(\frac{1 \text{ mole O}}{16.00 \text{ g O}}\right) = 0.0713 \text{ moles oxygen}$$

Then calculate the moles of X in the oxygen compound:

$$(0.0713 \text{ moles O})\left(\frac{1 \text{ mole X}}{2 \text{ moles O}}\right) = 0.0356 \text{ moles X}$$

Finally, calculate the molar mass of X:

$$\frac{1.00 \text{ g X}}{0.0356 \text{ moles X}} = 28.1 \text{ g / mole X}$$

Note that X is Si, atomic number 14, and that its oxygen compound is SiO_2.

(b) The same number of moles of X are also combined with Y:

$$(0.0356 \text{ moles X})\left(\frac{4 \text{ moles Y}}{1 \text{ moles X}}\right) = 0.142 \text{ moles Y}$$

The atomic mass of Y is thus:

$$\frac{5.07 \text{ g Y}}{0.142 \text{ moles Y}} = 35.7 \text{ g / mole Y}$$

Note that Y is Cl, atomic number 17, and that its compound with X is $SiCl_4$.

2.139 (a) $3H_2SO_4 + 2Al(OH)_3 \rightarrow Al_2(SO_4)_3 + 6H_2O$

(b) $H_2SO_4 + Ca(OH)_2 \rightarrow CaSO_4 + 2H_2O$

2.141 (a)   hypochlorous acid and sodium hypochlorite; NaOCl
      (b)   iodous acid and sodium iodite; $NaIO_2$
      (c)   bromic acid and sodium bromate; $NaBrO_3$
      (d)   perchloric acid and sodium perchlorate; $NaClO_4$

CHAPTER THREE
Review Exercises

3.4   # atoms = $(1.5 \text{ mol})(6.02 \times 10^{23} \text{ atoms/mol}) = 9.03 \times 10^{23}$ atoms

      # g = $(1.5 \text{ mol})(12 \text{ g/mol}) = 18$ g

3.5 (a)   1 S, 2 O
     (b)   2 As, 3 O
     (c)   1 Pb, 6 N
     (d)   2 Al, 6 Cl

3.8   # mol V = $(0.565 \text{ mol O})\left(\dfrac{2 \text{ mol V}}{5 \text{ mol O}}\right) = 0.226$ mol V

3.12   # atoms = $(1.000 \times 10^{-12} \text{ g Pb})\left(\dfrac{1 \text{ mol Pb}}{207.2 \text{ g Pb}}\right)\left(\dfrac{6.022 \times 10^{23} \text{ atoms Pb}}{1 \text{ mol Pb}}\right)$

         = 2.906×10^{9} atoms Pb

3.14 Note: all masses are in g/mole

     (a) $NaHCO_3$ = 1Na + 1H + 1C + 3O
                    = (22.99) + (1.01) + (12.01) + (3 × 16.00)
                    = 84.01 g/mole
     (b) $K_2Cr_2O_7$ = 2K + 2Cr + 7O
                    = (2 × 39.10) + (2 × 52.00) + (7 × 16.00)
                    = 294.20 g/mole
     (c) $(NH_4)_2CO_3$ = 2N + 8H + C + 3O
                    = (2 × 14.01) + (8 × 1.01) + (12.01) + (3 × 16.0)
                    = 96.11 g/mole
     (d) $Al_2(SO_4)_3$ = 2Al + 3S + 12O
                    = (2 × 26.98) + (3 × 32.07) + (12 × 16.00)
                    = 342.17 g/mole
     (e) $CuSO_4 \cdot 5H_2O$ = 1Cu + 1S + 9O + 10H
                    = 63.55 + 32.07 + (9 × 16.00) + (10 × 1.01)
                    = 249.72 g/mole

3.17 (a) # g = $(1.25 \text{ mol Ca})(40.078 \text{ g Ca/1 mole Ca}) = 50.1$ g Ca
     (b) # g = $(0.625 \text{ mol Fe})(55.847 \text{ g Fe/1 mole Ca}) = 34.9$ g Fe
     (c) # g = $(0.600 \text{ mol } C_4H_{10})(58.12 \text{ g } C_4H_{10}/1 \text{ mole } C_4H_{10}) = 34.9$ g C_4H_{10}
     (d) # g = $(1.45 \text{ mol } (NH_4)_2CO_3)(96.11 \text{ g/mole}) = 139$ g $(NH_4)_2CO_3$
     (e) # g = $(2.15 \times 10^{-3} \text{ mol KMnO}_4)(158.0 \text{ g/mole}) = 0.340$ g $KMnO_4$

3.19 (a) # moles $CaCO_3$ = $(21.5 \text{ g } CaCO_3)\left(\dfrac{1 \text{ mole } CaCO_3}{100.09 \text{ g } CaCO_3}\right)$ = 0.215 moles $CaCO_3$

(b) # moles NH_3 = $(1.56 \text{ g } NH_3)\left(\dfrac{1 \text{ mole } NH_3}{17.03 \text{ g } NH_3}\right)$ = 9.16×10^{-2} moles NH_3

(c) # moles $Sr(NO_3)_2$ = $(16.8 \text{ g } Sr(NO_3)_2)\left(\dfrac{1 \text{ mole } Sr(NO_3)_2}{211.6 \text{ g } Sr(NO_3)_2}\right)$

= 7.94×10^{-2} moles $Sr(NO_3)_2$

(d) # moles Na_2CrO_4 = $(6.98 \times 10^{-6} \text{ g } Na_2CrO_4)\left(\dfrac{1 \text{ mole } Na_2CrO_4}{162.0 \text{ g } Na_2CrO_4}\right)$

= 4.31×10^{-8} moles Na_2CrO_4

3.21 (a) # g C = $(0.200 \text{ moles } Na_2CO_3)\left(\dfrac{1 \text{ mole C}}{1 \text{ mole } Na_2CO_3}\right)\left(\dfrac{12.011 \text{ g C}}{1 \text{ mole C}}\right)$ = 2.40 g C

(b) # g C = $(0.200 \text{ moles } C_3H_8)\left(\dfrac{3 \text{ mole C}}{1 \text{ mole } C_3H_8}\right)\left(\dfrac{12.011 \text{ g C}}{1 \text{ mole C}}\right)$ = 7.21 g C

(c) # g C = $(25.0 \text{ g } Fe_2(CO_3)_3)\left(\dfrac{1 \text{ mole } Fe_2(CO_3)_3}{291.7 \text{ g } Fe_2(CO_3)_3}\right)$

$\times \left(\dfrac{3 \text{ mole C}}{1 \text{ mole } Fe_2(CO_3)_3}\right)\left(\dfrac{12.011 \text{ g C}}{1 \text{ mole C}}\right)$

= 3.09 g C

(d) # g C = $(14.5 \text{ g } CO_2)\left(\dfrac{1 \text{ mole } CO_2}{44.01 \text{ g } CO_2}\right)\left(\dfrac{1 \text{ mole C}}{1 \text{ mole } CO_2}\right)\left(\dfrac{12.011 \text{ g C}}{1 \text{ mole C}}\right)$

= 3.96 g C

(e) # g C = $(25.0 \text{ g } C_6H_{14})\left(\dfrac{1 \text{ mole } C_6H_{14}}{86.18 \text{ g } C_6H_{14}}\right)\left(\dfrac{6 \text{ mole C}}{1 \text{ mole } C_6H_{14}}\right)\left(\dfrac{12.011 \text{ g C}}{1 \text{ mole C}}\right)$

= 20.9 g C

3.24 The formula CaC_2 indicates that there is 1 mole of Ca for every 2 moles of C. Therefore, if there are 0.150 moles of C there must be 0.0750 moles of Ca.

g Ca = $(0.075 \text{ mol Ca})\left(\dfrac{40.078 \text{ g Ca}}{1 \text{ mole Ca}}\right)$ = 3.01 g Ca

3.26   # mol N $= (0.650 \text{ mol (NH}_4)_2\text{CO}_3)\left(\dfrac{2 \text{ moles N}}{1 \text{ mole (NH}_4)_2\text{CO}_3}\right) = 1.30 \text{ moles N}$

$$\text{\# g (NH}_4)_2\text{CO}_3 = (0.650 \text{ mol (NH}_4)_2\text{CO}_3)\left(\dfrac{96.09 \text{ g (NH}_4)_2\text{CO}_3}{1 \text{ mole (NH}_4)_2\text{CO}_3}\right)$$

$$= 62.5 \text{ g (NH}_4)_2\text{CO}_3$$

3.28

$$\text{\# mol NaOH} = (13 \text{ oz NaOH})\left(\dfrac{28.4 \text{ g}}{1 \text{ oz}}\right)\left(\dfrac{1 \text{ mol NaOH}}{40.0 \text{ g NaOH}}\right) = 9.2 \text{ mol NaOH}$$

$$\text{\# mol tallow} = (6.0 \text{ lb tallow})\left(\dfrac{453.6 \text{ g}}{1 \text{ lb}}\right)\left(\dfrac{1 \text{ mol tallow}}{891.7 \text{ g tallow}}\right) = 3.1 \text{ mol tallow}$$

$$\text{mole ratio} = \dfrac{9.2 \text{ mol NaOH}}{3.1 \text{ mol tallow}} = 3.0$$

3.30   # mol $NH_3 = (34 \times 10^9 \text{ lb NH}_3)\left(\dfrac{453.6 \text{ g}}{1 \text{ lb}}\right)\left(\dfrac{1 \text{ mol NH}_3}{17.03 \text{ g NH}_3}\right) = 9.1 \times 10^{11} \text{ moles NH}_3$

3.32   (a)   # g O $= (24.00 \text{ g S})\left(\dfrac{1 \text{ mole S}}{32.066 \text{ g S}}\right)\left(\dfrac{2 \text{ moles O}}{1 \text{ mole S}}\right)\left(\dfrac{16.00 \text{ g O}}{1 \text{ mole O}}\right) = 23.95 \text{ g O}$

     (b)   # g O $= (24.00 \text{ g S})\left(\dfrac{1 \text{ mole S}}{32.066 \text{ g S}}\right)\left(\dfrac{3 \text{ moles O}}{1 \text{ mole S}}\right)\left(\dfrac{16.00 \text{ g O}}{1 \text{ mole O}}\right) = 35.93 \text{ g O}$

3.34   # g O $= (12.64 \text{ g S})\left(\dfrac{1 \text{ mole S}}{32.066 \text{ g S}}\right)\left(\dfrac{4 \text{ mole O}}{1 \text{ mole S}}\right)\left(\dfrac{15.999 \text{ g O}}{1 \text{ mole O}}\right) = 25.23 \text{ g O}$

       # g H $= (12.64 \text{ g S})\left(\dfrac{1 \text{ mole S}}{32.066 \text{ g S}}\right)\left(\dfrac{2 \text{ mole H}}{1 \text{ mole S}}\right)\left(\dfrac{1.008 \text{ g H}}{1 \text{ mole H}}\right) = 0.7947 \text{ g H}$

3.36   (a)   Since only two elements are present, oxygen constitutes the difference between 100 % and the percentage by weight of C: 100.00 % – 27.29 % = 72.71 % O

     (b)   # g C $= (50.00 \text{ g CO}_2)\left(\dfrac{1 \text{ mol CO}_2}{44.011 \text{ g CO}_2}\right)\left(\dfrac{1 \text{ mole C}}{1 \text{ mole CO}_2}\right)\left(\dfrac{12.011 \text{ g C}}{1 \text{ mole C}}\right)$

$$= 13.65 \text{ g C}$$

3.38   For NH_3, the molar mass (1N + 3H) equals 17.0 g/mole, and the % by weight N is given

by: $\dfrac{14.0 \text{ g N}}{17.0 \text{ NH}_3} \times 100\% = 82.4\%.$

For $CO(NH_2)_2$, the molar mass (2N + 4H + 1C + 1O) equals 60.1 g/mole, and the % by weight N is given by: $\dfrac{28.0 \text{ g N}}{60.1 \text{ CO(NH}_2)_2} \times 100\% = 46.6\%.$

3.40   For $C_{17}H_{25}N$, the molar mass (17C + 25H + 1N) equals 243.39 g/mole, and the three theoretical values for % by weight are calculated as follows:

$$\% \text{ C} = \frac{204.2 \text{ g C}}{243.4 \text{ g C}_{17}\text{H}_{25}\text{N}} \times 100\% = 83.89\%$$

$$\% \text{ H} = \frac{25.20 \text{ g H}}{243.4 \text{ g C}_{17}\text{H}_{25}\text{N}} \times 100\% = 10.35\%$$

$$\% \text{ N} = \frac{14.01 \text{ g N}}{243.4 \text{ g C}_{17}\text{H}_{25}\text{N}} \times 100\% = 5.756\%$$

These data are consistent with the experimental values cited in the problem.

3.44   The molecular formula is some integer multiple of the empirical formula. This means that we can divide the molecular formula by the largest possible whole number that gives an integer ratio among the atoms in the empirical formula.
(a) SCl   (b) CH_2O (c) C_2H_5 (d) AsO_3 (e) HO

3.46   From the information provided, we can determine the mass of oxygen as the difference between the total mass and the mass of phosphorus:

\# g O = 0.3148 g compound − 0.1774 g P = 0.1374 g O

To determine the empirical formula first convert the two masses to a number of moles.

$$\# \text{ moles P} = \left(0.1774 \text{ g P}\right)\left(\frac{1 \text{ mole P}}{30.974 \text{ g P}}\right) = 5.727 \times 10^{-3} \text{ moles P}$$

$$\# \text{ moles O} = \left(0.1374 \text{ g O}\right)\left(\frac{1 \text{ mole O}}{15.999 \text{ g O}}\right) = 8.588 \times 10^{-3} \text{ moles O}$$

Now, divide each of these values by the smaller quantity to determine the simplest mole ratio between the two elements:

for P: 5.727×10^{-3} moles$/5.727 \times 10^{-3}$ moles $= 1.000$

for O: 8.588×10^{-3} moles$/5.727 \times 10^{-3}$ moles $= 1.500$

The relative mole rations are then 1:1.5 or, since chemical formulas use integer values, 2:3. The empirical formula is thus P_2O_3.

3.48   From the information provided, the mass of sulfur is the difference between the total mass and the mass of antimony:

g S $= 0.6662$ g compound $- 0.4017$ g Sb $= 0.2645$ g S

To determine the empirical formula first convert the two masses to a number of moles.

$$\text{\# moles S} = \left(0.2645 \text{ g S}\right)\left(\frac{1 \text{ mole S}}{32.066 \text{ g S}}\right) = 8.249 \times 10^{-3} \text{ moles S}$$

$$\text{\# moles Sb} = \left(0.4017 \text{ g Sb}\right)\left(\frac{1 \text{ mole Sb}}{121.76 \text{ g Sb}}\right) = 3.299 \times 10^{-3} \text{ moles Sb}$$

Now, divide each of these values by the smaller quantity to determine the simplest mole ratio between the two elements:

For Sb: $3.299 \times 10^{-3}/3.299 \times 10^{-3} = 1.000$ mol Sb

For S: $8.249 \times 10^{-3}/3.299 \times 10^{-3} = 2.500$ mol S

Hence the empirical formula is Sb_2S_5, and the empirical mass is $(2 \times Sb) + (5 \times S) = 403.85$ g/mol. Since the molecular mass reported in the problem is the same as the calculated empirical mass, the empirical formula is the same as the molecular formula.

3.50   To solve this problem we will assume that we have a 100 g sample. This implies that we have 77.26 g Hg, 9.25 g C, 1.17 g H and 12.32 g O. The amount of oxygen was determined by subtracting the total amounts of the other three elements from the total assumed mass of 100 g. Convert each of these masses into a number of moles:

$$\# \text{ moles Hg} = (77.26 \text{ g Hg})\left(\frac{1 \text{ mole Hg}}{200.59 \text{ g Hg}}\right) = 0.3852 \text{ moles Hg}$$

$$\# \text{ moles C} = (9.25 \text{ g C})\left(\frac{1 \text{ mole C}}{12.011 \text{ g C}}\right) = 0.770 \text{ moles C}$$

$$\# \text{ moles H} = (1.17 \text{ g H})\left(\frac{1 \text{ mole H}}{1.008 \text{ g H}}\right) = 1.16 \text{ moles H}$$

$$\# \text{ moles O} = (12.32 \text{ g O})\left(\frac{1 \text{ mole O}}{15.999 \text{ g O}}\right) = 0.7700 \text{ moles O}$$

The relative mole amounts are determined as follows:

for Hg, $0.3852 \div 0.3852 = 1.000$
for C, $0.770 \div 0.3852 = 2.00$
for H, $1.16 \div 0.3852 = 3.01$
for O, $0.7700 \div 0.3852 = 1.999$

and the empirical formula is $HgC_2H_3O_2$. The empirical formula weight is 260 g/mole, which must be multiplied by 2 in order to obtain the molecular weight. This means that the molecular formula is twice the empirical formula, or $Hg_2C_4H_6O_4$.

3.52 (a) We follow the following procedure: (i) convert to moles, (ii) determine the simplest mole ratio, (iii) convert mole ratio values to integers if necessary.

$$\# \text{ moles Na} = (0.4681 \text{ g Na})\left(\frac{1 \text{ mole Na}}{23.000 \text{ g Na}}\right) = 0.02035 \text{ moles Na}$$

$$\# \text{ moles O} = (0.3258 \text{ g O})\left(\frac{1 \text{ mole O}}{15.999 \text{ g O}}\right) = 0.02036 \text{ moles O}$$

Since these mole amounts are clearly in a ratio of 1 to 1, the empirical formula is NaO.

(b) $\dfrac{0.4681 \text{ g Na}}{0.3258 \text{ g O}} = \dfrac{X \text{ g Na}}{1.000 \text{ g O}}$ solving gives $X = 1.437 \text{ g Na}$

(c) Any integer multiple or any whole number fraction of the value 1.437 g will be a possible amount of Na to combine with 1.000 g of O to make a new compound of sodium and oxygen; $2(1.437 \text{ g Na}) = 2.874 \text{ g Na}$, $3(1.437 \text{ g Na}) = 4.311 \text{ g Na}$, $(1/2)(1.437 \text{ g Na}) = 0.7185 \text{ g Na}$, $(1/3)(1.437 \text{ g Na}) = 0.4790 \text{ g Na}$, etc...

(d) $\dfrac{1.145 \text{ g Na}}{0.3983 \text{ g O}} = \dfrac{X \text{ g Na}}{1.000 \text{ g O}}$ solving gives $X = 2.875$ g Na

Notice that in this different compound of sodium and oxygen, the mass ratio of the two elements is different from that found in part (b) to this question, as is guaranteed by the Law of Definite Proportions.

(e) In part (c) we determined that 1.436 g Na combines with 1.000 g O. In part (d) we found 2.875 g Na combines with 1.000 g O. The ratio of these two masses is:

$$\frac{1.437 \text{ g Na}}{2.875 \text{ g Na}} = \frac{1}{2}$$

Since this gives a ratio of small whole numbers, it is entirely consistent with the Law of Multiple Proportions.

(f) First convert the masses of Na and O to moles:

$$\# \text{ moles Na} = (1.145 \text{ g Na})\left(\frac{1 \text{ mole Na}}{23.000 \text{ g Na}}\right) = 0.04978 \text{ moles Na}$$

$$\# \text{ moles O} = (0.3983 \text{ g O})\left(\frac{1 \text{ mole O}}{15.999 \text{ g O}}\right) = 0.02490 \text{ moles O}$$

Now, determine the simplest mole ratio:

4.978×10^{-2} mole $\div$ 2.490×10^{-2} mole $= 1.999$ relative moles of Na

$2.490 \times 10^{-2} \div 2.490 \times 10^{-2} = 1.000$ relative moles of O

and the empirical formula is seen to be Na_2O.

3.54 First, we determine the number of grams of chlorine in the original sample:

$$\# \text{ g Cl} = (0.3383 \text{ g AgCl})\left(\frac{1 \text{ mole AgCl}}{143.32 \text{ g AgCl}}\right)\left(\frac{1 \text{ mole Cl}}{1 \text{ mole AgCl}}\right)\left(\frac{35.453 \text{ g Cl}}{1 \text{ mole Cl}}\right)$$

$$= 0.08369 \text{ g Cl}$$

The mass of Cr in the original sample is thus $0.1246 - 0.08369$ g $= 0.0409$ g Cr. Converting to moles, we have:

for Cl: 0.08369 g $\div 35.453$ g/mol $= 2.361 \times 10^{-3}$ mol Cl

for Cr: 0.0409 g Cr $\div 52.00$ g/mol $= 7.87 \times 10^{-4}$ mol Cr

The relative mole amounts are:

for Cl: $2.361 \times 10^{-3}/7.87 \times 10^{-4} = 3.00$

for Cr: $7.87 \times 10^{-4}/7.87 \times 10^{-4} = 1.00$

The empirical formula is thus $CrCl_3$.

3.56   This type of combustion analysis takes advantage of the fact that the entire amount of carbon in the original sample appears as CO_2 among the products. Hence the mass of carbon in the original sample must be equal to the mass of carbon that is found in the CO_2.

$$\# \text{ g C} = (19.73 \times 10^{-3} \text{ g CO}_2)\left(\frac{1 \text{ mole CO}_2}{44.01 \text{ g CO}_2}\right)\left(\frac{1 \text{ mole C}}{1 \text{ mole CO}_2}\right)\left(\frac{12.011 \text{ g C}}{1 \text{ mole C}}\right)$$

$$= 5.385 \times 10^{-3} \text{ g C}$$

Similarly, the entire mass of hydrogen that was present in the original sample ends up in the products as H_2O:

$$\# \text{ g H} = (6.391 \times 10^{-3} \text{ g H}_2\text{O})\left(\frac{1 \text{ mole H}_2\text{O}}{18.02 \text{ g H}_2\text{O}}\right)\left(\frac{2 \text{ mole H}}{1 \text{ mole H}_2\text{O}}\right)\left(\frac{1.008 \text{ g H}}{1 \text{ mole H}}\right)$$

$$= 7.150 \times 10^{-4} \text{ g H}$$

The mass of oxygen is determined by subtracting the mass due to C and H from the total mass: 6.853 mg total $- (5.385$ mg C $+ 0.7150$ mg H$) = 0.753$ mg O. Now, convert these masses to a number of moles:

$$\# \text{ moles C} = \left(5.385 \times 10^{-3} \text{ g C}\right)\left(\frac{1 \text{ mole C}}{12.011 \text{ g C}}\right) = 4.483 \times 10^{-4} \text{ moles C}$$

$$\# \text{ moles H} = \left(7.150 \times 10^{-4} \text{ g H}\right)\left(\frac{1 \text{ mole H}}{1.0079 \text{ g H}}\right) = 7.094 \times 10^{-4} \text{ moles H}$$

$$\# \text{ moles O} = \left(7.53 \times 10^{-4} \text{ g O}\right)\left(\frac{1 \text{ mole O}}{15.999 \text{ g O}}\right) = 4.71 \times 10^{-5} \text{ moles O}$$

The relative mole amounts are:

$$\text{for C, } 4.483 \times 10^{-4} \div 4.71 \times 10^{-5} = 9.52$$
$$\text{for H, } 7.094 \times 10^{-4} \div 4.71 \times 10^{-5} = 15.1$$
$$\text{for O, } 4.71 \times 10^{-5} \div 4.71 \times 10^{-5} = 1.00$$

The relative mole amounts are not nice whole numbers as we would like. However, we see that if we double the relative number of moles of each compound, there are approximately 19 moles of C, 30 moles of H and 2 moles of O. If we assume these numbers are correct, the empirical formula is $C_{19}H_{30}O_2$, for which the formula weight is 290 g/mole. Since the molar mass is equal to the mass of this empirical formula, the molecular formula is the same as this empirical formula, $C_{19}H_{30}O_2$ and we have performed the analysis correctly.

In most problems where we attempt to determine an empirical formula, the relative mole amounts should work out to give a "nice" set of values for the formula. Rarely will a problem be designed that gives very odd coefficients. With experience and practice, you will recognize when a set of values is reasonable.

3.59   14 moles of Fe

3.64   (a) $Ca(OH)_2 + 2HCl \rightarrow CaCl_2 + 2H_2O$

(b) $2AgNO_3 + CaCl_2 \rightarrow Ca(NO_3)_2 + 2AgCl$

(c) $2Fe_2O_3 + 3C \rightarrow 4Fe + 3CO_2$

(d) $2NaHCO_3 + H_2SO_4 \rightarrow Na_2SO_4 + 2H_2O + 2CO_2$

(e) $2C_4H_{10} + 13O_2 \rightarrow 8CO_2 + 10H_2O$

3.69   (a) $\# \text{ moles Na}_2S_2O_3 = (0.12 \text{ moles Cl}_2)\left(\dfrac{1 \text{ mole Na}_2S_2O_3}{4 \text{ mole Cl}_2}\right)$
$$= 0.030 \text{ moles Na}_2S_2O_3$$

(b) $\# \text{ moles HCl} = (0.12 \text{ moles Cl}_2)\left(\dfrac{8 \text{ mole HCl}}{4 \text{ mole Cl}_2}\right) = 0.24 \text{ moles HCl}$

(c) $\# \text{ moles H}_2O = (0.12 \text{ moles Cl}_2)\left(\dfrac{5 \text{ mole H}_2O}{4 \text{ mole Cl}_2}\right) = 0.15 \text{ moles H}_2O$

3.71   (a) $4P + 5O_2 \rightarrow P_4O_{10}$

      (b)   # moles O_2 = (0.221 moles P)$\left(\dfrac{5 \text{ mole } O_2}{4 \text{ mole } P}\right)$ = 0.276 moles O_2

      (c)   # moles P_4O_{10} = (0.250 moles O_2)$\left(\dfrac{1 \text{ mole } P_4O_{10}}{5 \text{ mole } O_2}\right)$ = 0.0500 moles P_4O_{10}

      (d)   # moles P = (0.114 moles O_2)$\left(\dfrac{4 \text{ mole } P}{5 \text{ mole } O_2}\right)$ = 0.0912 moles P

3.73   (a) $Fe_2O_3 + 3H_2 \rightarrow 2Fe + 3H_2O$

      (b)   # moles Fe = (22 moles Fe_2O_3)$\left(\dfrac{2 \text{ mole } Fe}{1 \text{ mole } Fe_2O_3}\right)$ = 44 moles Fe

      (c)   # moles H_2 = (24 moles Fe)$\left(\dfrac{3 \text{ mole } H_2}{2 \text{ mole } Fe}\right)$ = 36 moles H_2

      (d)

$$\text{\# g } Fe_2O_3 = (95 \text{ moles } H_2O)\left(\frac{1 \text{ mole } Fe_2O_3}{3 \text{ moles } H_2O}\right)\left(\frac{159.69 \text{ g } Fe_2O_3}{1 \text{ mole } Fe_2O_3}\right)$$

$$= 5.1 \times 10^3 \text{ g } Fe_2O_3$$

3.75 $\text{\# moles of } HNO_3 = (11.45 \text{ g Cu})\left(\dfrac{1 \text{ mole Cu}}{63.546 \text{ g Cu}}\right)\left(\dfrac{8 \text{ moles } HNO_3}{3 \text{ moles Cu}}\right)$

$$= 0.4805 \text{ moles } HNO_3$$

3.77   (a)   If all of the Zn were to react, then the required amount of S would be:

$$\text{\# g S} = (30.0 \text{ g Zn})\left(\frac{1 \text{ mole Zn}}{65.39 \text{ g Zn}}\right)\left(\frac{1 \text{ mole S}}{1 \text{ mole Zn}}\right)\left(\frac{32.066 \text{ g S}}{1 \text{ mole S}}\right)$$

$$= 14.7 \text{ g S}$$

Similarly, the amount of Zn that would be required to react with all of the available S would be:

$$\text{\# g Zn} = (36.0 \text{ g S})\left(\frac{1 \text{ mole S}}{32.066 \text{ g S}}\right)\left(\frac{1 \text{ mole Zn}}{1 \text{ mole S}}\right)\left(\frac{65.39 \text{ g Zn}}{1 \text{ mole Zn}}\right)$$

$$= 73.4 \text{ g Zn}$$

Since only 30.0 g Zn are available, Zn is the limiting reactant.

(b)

$$\# \text{ g ZnS } = (30.0 \text{ g Zn}) \left(\frac{1 \text{ mole Zn}}{65.39 \text{ g Zn}} \right) \left(\frac{1 \text{ mole ZnS}}{1 \text{ mole Zn}} \right) \left(\frac{97.46 \text{ g ZnS}}{1 \text{ mole ZnS}} \right)$$

$$= 44.7 \text{ g ZnS}$$

(c)   From part (a) we know that 14.7 g S will react completely with all of the 30.0 g Zn. The unreacted amount of S is therefore: 36.0 g S total – 14.7 g S reacted = 21.3 g S unused.

3.79   (a)   First calculate the number of moles of water that are needed to react completely with the given amount of PCl_5:

$$\# \text{ moles H}_2\text{O } = (0.360 \text{ moles PCl}_5) \left(\frac{4 \text{ mole H}_2\text{O}}{1 \text{ moles PCl}_5} \right) = 1.44 \text{ moles H}_2\text{O}$$

Since this is less than the amount of water that is supplied, the limiting reactant must be PCl_5. This can be confirmed by the following calculation:

$$\# \text{ moles PCl}_5 = (2.88 \text{ moles H}_2\text{O}) \left(\frac{1 \text{ moles PCl}_5}{4 \text{ mole H}_2\text{O}} \right) = 0.720 \text{ moles PCl}_5$$

which also demonstrates that the limiting reactant is PCl_5.

(b)

$$\# \text{ moles H}_3\text{PO}_4 = (0.360 \text{ moles PCl}_5) \left(\frac{1 \text{ mole H}_3\text{PO}_4}{1 \text{ moles PCl}_5} \right)$$

$$= 0.360 \text{ moles H}_3\text{PO}_4$$

$$\# \text{ moles HCl } = (0.360 \text{ moles PCl}_5) \left(\frac{5 \text{ mole HCl}}{1 \text{ moles PCl}_5} \right) = 1.80 \text{ moles HCl}$$

3.82   (a)

$$\# \text{ g H}_2\text{SO}_4 = (25.00 \text{ g AlCl}_3) \left(\frac{1 \text{ mole AlCl}_3}{133.34 \text{ g AlCl}_3} \right)$$

$$\times \left(\frac{3 \text{ mole H}_2\text{SO}_4}{2 \text{ mole AlCl}_3} \right) \left(\frac{98.08 \text{ g H}_2\text{SO}_4}{1 \text{ mole H}_2\text{SO}_4} \right)$$

$$= 27.58 \text{ g H}_2\text{SO}_4$$

(b)   First determine the theoretical yield:

$$\# \text{ g Al}_2(SO_4)_3 = (25.00 \text{ g AlCl}_3)\left(\frac{1 \text{ mole AlCl}_3}{133.34 \text{ g AlCl}_3}\right)$$

$$\times \left(\frac{1 \text{ mole Al}_2(SO_4)_3}{2 \text{ mole AlCl}_3}\right)\left(\frac{342.15 \text{ g Al}_2(SO_4)_3}{1 \text{ mole Al}_2(SO_4)_3}\right)$$

$$= 32.07 \text{ g Al}_2(SO_4)_3$$

Then calculate a % yield:

$$\% \text{ yield} = \frac{\text{actual yield}}{\text{theoretical yield}} \times 100 = \frac{28.36 \text{ g}}{32.07 \text{ g}} \times 100 = 88.43\%$$

3.84   If the yield for this reaction is only 71 % and we need to have 11.5 g of product, we will attempt to make 16 g of product. This is determined by dividing the actual yield by the percent yield. Recall that; $\% \text{ yield} = \dfrac{\text{actual yield}}{\text{theoretical yield}} \times 100$. If we rearrange this equation we can see that $\text{theoretical yield} = \dfrac{\text{actual yield}}{\% \text{ yield}} \times 100$. Substituting the values from this problem gives the 16 g of product mentioned above.

$$\# \text{ g C}_7\text{H}_8 = (16 \text{ g KC}_7\text{H}_5\text{O}_2)\left(\frac{1 \text{ mole KC}_7\text{H}_5\text{O}_2}{160.21 \text{ g KC}_7\text{H}_5\text{O}_2}\right)$$

$$\times \left(\frac{1 \text{ mole C}_7\text{H}_8}{1 \text{ mole KC}_7\text{H}_5\text{O}_2}\right)\left(\frac{92.14 \text{ g C}_7\text{H}_8}{1 \text{ mole C}_7\text{H}_8}\right)$$

$$= 9.2 \text{ g C}_7\text{H}_8$$

3.86

$$\# \text{ g (NH}_2)_2\text{CO} = (6.00 \text{ g N})\left(\frac{1 \text{ mole N}}{14.007 \text{ g N}}\right)\left(\frac{1 \text{ mole (NH}_2)_2\text{CO}}{2 \text{ moles N}}\right)\left(\frac{60.06 \text{ g (NH}_2)_2\text{CO}}{1 \text{ mole (NH}_2)_2\text{CO}}\right)$$

$$= 12.9 \text{ g (NH}_2)_2\text{CO}$$

3.88   (a) $Mg(OH)_2 + 2HBr \rightarrow MgBr_2 + 2H_2O$

(b) $2HCl + Ca(OH)_2 \rightarrow CaCl_2 + 2H_2O$

(c) $Al_2O_3 + 3H_2SO_4 \rightarrow Al_2(SO_4)_3 + 3H_2O$

(d) $2KHCO_3 + H_3PO_4 \rightarrow K_2HPO_4 + 2H_2O + 2CO_2$

(e) $C_9H_{20} + 14O_2 \rightarrow 9CO_2 + 10H_2O$

3.90   First, determine the amount of oxygen in the sample by subtracting the masses of the other elements from the total mass: $0.6216 \text{ g} - (0.1735 \text{ g C} + 0.01455 \text{ g H} + 0.2024 \text{ g N}) = 0.2312 \text{ g O}$. Now, convert these masses into a number of moles for each element:

$$\# \text{ moles C} = (0.1735 \text{ g C})\left(\frac{1 \text{ mole C}}{12.011 \text{ g C}}\right) = 1.445 \times 10^{-2} \text{ moles C}$$

$$\# \text{ moles H} = (0.01455 \text{ g H})\left(\frac{1 \text{ mole H}}{1.0079 \text{ g H}}\right) = 1.444 \times 10^{-2} \text{ moles H}$$

$$\# \text{ moles N} = (0.2024 \text{ g N})\left(\frac{1 \text{ mole N}}{14.007 \text{ g N}}\right) = 1.445 \times 10^{-2} \text{ moles N}$$

$$\# \text{ moles O} = (0.2312 \text{ g O})\left(\frac{1 \text{ mole O}}{15.999 \text{ g O}}\right) = 1.445 \times 10^{-2} \text{ moles O}$$

These are clearly all the same mole amounts, and we deduce that the empirical formula is CHNO, which has a formula weight of 43. It can be seen that the number 43 must be multiplied by the integer 3 in order to obtain the molar mass ($3 \times 43 = 129$), and this means that the empirical formula should similarly be multiplied by 3 in order to arrive at the molecular formula, $C_3H_3N_3O_3$.

CHAPTER FOUR
Review Exercises

4.6　(a)　$LiCl(s) \rightarrow Li^+(aq) + Cl^-(aq)$

　　(b)　$BaCl_2(s) \rightarrow Ba^{2+}(aq) + 2Cl^-(aq)$

　　(c)　$Al(C_2H_3O_2)_3(s) \rightarrow Al^{3+}(aq) + 3C_2H_3O_2^-(aq)$

　　(d)　$(NH_4)_2CO_3(s) \rightarrow 2NH_4^+(aq) + CO_3^{2-}(aq)$

　　(e)　$FeCl_3(s) \rightarrow Fe^{3+}(aq) + 3Cl^-(aq)$

4.11　This is an ionization reaction: $HClO_4(\ell) + H_2O(\ell) \rightarrow H_3O^+(aq) + ClO_4^-(aq)$

4.13　a c and d are all nonmetal oxides and would react with water to form molecular acids.

4.16　$HNO_2(aq) + H_2O(\ell) \rightleftharpoons H_3O^+(aq) + NO_2^-(aq)$

4.19　$HClO_3(aq) + H_2O(\ell) \rightarrow H_3O^+(aq) + ClO_3^-(aq)$

4.21　$H_3PO_4(aq) + H_2O(\ell) \rightleftharpoons H_3O^+(aq) + H_2PO_4^-(aq)$

　　　$H_2PO_4^-(aq) + H_2O(\ell) \rightleftharpoons H_3O^+(aq) + HPO_4^{2-}(aq)$

　　　$HPO_4^{2-}(aq) + H_2O(\ell) \rightleftharpoons H_3O^+(aq) + PO_4^{3-}(aq)$

4.25　(a)　molecular: $Ca(OH)_2(aq) + 2HNO_3(aq) \rightarrow Ca(NO_3)_2(aq) + 2H_2O$

　　　　　ionic: $Ca^{2+}(aq) + 2OH^-(aq) + 2H^+(aq) + 2NO_3^-(aq) \rightarrow$
　　　　　　　　　　$Ca^{2+}(aq) + 2NO_3^-(aq) + 2H_2O$

　　　　　net: $H^+(aq) + OH^-(aq) \rightarrow H_2O$

　　(b)　molecular: $Al_2O_3(s) + 6HCl(aq) \rightarrow 2AlCl_3(aq) + 3H_2O$

　　　　　ionic: $Al_2O_3(s) + 6H^+(aq) + 6Cl^-(aq) \rightarrow$
　　　　　　　　　　$2Al^{3+}(aq) + 6Cl^-(aq) + 3H_2O$

　　　　　net: $Al_2O_3(s) + 6H^+(aq) \rightarrow 2Al^{3+}(aq) + 3H_2O$

(c) molecular: $Zn(OH)_2(s) + H_2SO_4(aq) \rightarrow ZnSO_4(aq) + 2H_2O$

ionic: $Zn(OH)_2(s) + 2H^+(aq) + SO_4^{2-}(aq) \rightarrow$
$$Zn^{2+}(aq) + SO_4^{2-}(aq) + 2H_2O$$

net: $Zn(OH)_2(s) + 2H^+(aq) \rightarrow Zn^{2+}(aq) + 2H_2O$

4.28 (a) ionic: $2NH_4^+(aq) + CO_3^{2-}(aq) + Mg^{2+}(aq) + 2Cl^-(aq) \rightarrow$
$$2NH_4^+(aq) + 2Cl^-(aq) + MgCO_3(s)$$

net: $Mg^{2+}(aq) + CO_3^{2-}(aq) \rightarrow MgCO_3(s)$

(b) ionic: $Cu^{2+}(aq) + 2Cl^-(aq) + 2Na^+(aq) + 2OH^-(aq) \rightarrow$
$$Cu(OH)_2(s) + 2Na^+(aq) + 2Cl^-(aq)$$

net: $Cu^{2+}(aq) + 2OH^-(aq) \rightarrow Cu(OH)_2(s)$

(c) ionic: $3Fe^{2+}(aq) + 3SO_4^{2-}(aq) + 6Na^+(aq) + 2PO_4^{3-}(aq) \rightarrow$
$$Fe_3(PO_4)_2(s) + 6Na^+(aq) + 3SO_4^{2-}(aq)$$

net: $3Fe^{2+}(aq) + 2PO_4^{3-}(aq) \rightarrow Fe_3(PO_4)_2(s)$

(d) ionic: $2Ag^+(aq) + 2C_2H_3O_2^-(aq) + Ni^{2+}(aq) + 2Cl^-(aq) \rightarrow$
$$2AgCl(s) + Ni^{2+}(aq) + 2C_2H_3O_2^-(aq)$$

net: $2Ag^+(aq) + 2Cl^-(aq) \rightarrow 2AgCl(s)$

4.30 $Cu^{2+}(aq) + S^{2-}(aq) \rightarrow CuS(s)$

4.32 molecular: $AgNO_3(aq) + NaBr(aq) \rightarrow AgBr(s) + NaNO_3(aq)$

ionic: $Ag^+(aq) + NO_3^-(aq) + Na^+(aq) + Br^-(aq) \rightarrow AgBr(s) + Na^+(aq) + NO_3^-(aq)$

net: $Ag^+(aq) + Br^-(aq) \rightarrow AgBr(s)$

4.34 (a) $HCl(aq) + NaHCO_3(aq) \rightarrow NaCl(aq) + H_2O(\ell) + CO_2(g)$

(b) $2HCl(aq) + Na_2S(aq) \rightarrow 2NaCl(aq) + H_2S(g)$

(c) $2HCl(aq) + K_2SO_3(aq) \rightarrow 2KCl(aq) + H_2O(\ell) + SO_2(g)$

4.36   (a) $2H^+(aq) + CO_3^{2-}(aq) \rightarrow H_2O(\ell) + CO_2(g)$

         (b) $NH_4^+(aq) + OH^-(aq) \rightarrow NH_3(aq) + H_2O(\ell)$

         (c) $H^+(aq) + HSO_3^-(aq) \rightarrow H_2O(\ell) + SO_2(g)$

4.37   The soluble ones are (a), (b), and (d).

4.38   The insoluble ones are (a), (d), and (f).

4.41   (a) $Na_2SO_3(aq) + Ba(NO_3)_2(aq) \rightarrow BaSO_3(s) + 2NaNO_3(aq)$

            ionic: $2Na^+(aq) + SO_3^{2-}(aq) + Ba^{2+}(aq) + 2NO_3^-(aq) \rightarrow$
$$BaSO_3(s) + 2Na^+(aq) + 2NO_3^-(aq)$$

            net: $Ba^{2+}(aq) + SO_3^{2-}(aq) \rightarrow BaSO_3(s)$

         (b) $K_2S(aq) + ZnCl_2(aq) \rightarrow ZnS(s) + 2KCl(aq)$

            ionic: $2K^+(aq) + S^{2-}(aq) + Zn^{2+}(aq) + 2Cl^-(aq) \rightarrow$
$$ZnS(s) + 2K^+(aq) + 2Cl^-(aq)$$

            net: $Zn^{2+}(aq) + S^{2-}(aq) \rightarrow ZnS(s)$

         (c) $2NH_4Br(aq) + Pb(C_2H_3O_2)_2(aq) \rightarrow 2NH_4C_2H_3O_2(aq) + PbBr_2(s)$

            ionic: $2NH_4^+(aq) + 2Br^-(aq) + Pb^{2+}(aq) + 2C_2H_3O_2^-(aq) \rightarrow$
$$2NH_4^+(aq) + 2C_2H_3O_2^-(aq) + PbBr_2(s)$$

            net: $Pb^{2+}(aq) + 2Br^-(aq) \rightarrow PbBr_2(s)$

         (d) $2NH_4ClO_4(aq) + Cu(NO_3)_2(aq) \rightarrow Cu(ClO_4)_2(aq) + 2NH_4NO_3(aq)$

            ionic: $2NH_4^+(aq) + 2ClO_4^-(aq) + Cu^{2+}(aq) + 2NO_3^-(aq) \rightarrow$
$$Cu^{2+}(aq) + 2ClO_4^-(aq) + 2NO_3^-(aq) + 2NH_4^+(aq)$$

            net: N.R.

4.43   (a) $3HNO_3(aq) + Cr(OH)_3(s) \rightarrow Cr(NO_3)_3(aq) + 3H_2O(\ell)$

            ionic: $3H^+(aq) + 3NO_3^-(aq) + Cr(OH)_3(s) \rightarrow$
$$Cr^{3+}(aq) + 3NO_3^-(aq) + 3H_2O(\ell)$$

            net: $3H^+(aq) + Cr(OH)_3(s) \rightarrow Cr^{3+}(aq) + 3H_2O(\ell)$

(b) $HClO_4(aq) + NaOH(aq) \rightarrow NaClO_4(aq) + H_2O(\ell)$

ionic: $H^+(aq) + ClO_4^-(aq) + Na^+(aq) + OH^-(aq) \rightarrow$

$$Na^+(aq) + ClO_4^-(aq) + H_2O(\ell)$$

net: $H^+(aq) + OH^-(aq) \rightarrow H_2O(\ell)$

(c) $Cu(OH)_2(s) + 2HC_2H_3O_2(aq) \rightarrow Cu(C_2H_3O_2)_2(aq) + 2H_2O(\ell)$

ionic: $Cu(OH)_2(s) + 2H^+ + 2C_2H_3O_2^-(aq) \rightarrow$

$$Cu^{2+}(aq) + 2C_2H_3O_2^-(aq) + 2H_2O(\ell)$$

net: $Cu(OH)_2(s) + 2H^+(aq) \rightarrow Cu^{2+}(aq) + 2H_2O(\ell)$

(d) $ZnO(s) + 2HBr(aq) \rightarrow ZnBr_2(aq) + H_2O(\ell)$

ionic: $ZnO(s) + 2H^+(aq) + 2Br^-(aq) \rightarrow Zn^{2+}(aq) + 2Br^-(aq) + H_2O(\ell)$

net: $ZnO(s) + 2H^+(aq) \rightarrow Zn^{2+}(aq) + H_2O(\ell)$

4.45   The electrical conductivity would decrease regularly, until one solution had neutralized the other, forming a nonelectrolyte:

$$Ba^{2+}(aq) + 2OH^-(aq) + 2H^+(aq) + SO_4^{2-}(aq) \rightarrow BaSO_4(s) + 2H_2O(\ell)$$

Once the point of neutralization had been reached, the addition of excess sulfuric acid would cause the conductivity to increase, because sulfuric acid is a strong electrolyte itself.

4.49   We need to choose a set of reactants that are both soluble and that react to yield only one solid product. Choose (b) and (d).

4.50   These reactions have the following "driving forces":
(a)   formation of insoluble $Cr(OH)_3$
(b)   formation of water, a weak electrolyte
(c)   formation of a gas, CO_2
(d)   formation of a weak electrolyte, $H_2C_2O_4$

4.52   The oxidation number of the Mg changes from 0 to +2. The oxidation number of the oxygen changes from 0 in O_2 to –2 in MgO. The magnesium is oxidized and the oxygen is reduced. Consequently, Mg is the reducing agent and the O_2 is the oxidizing agent.

4.55   Recall that the sum of the oxidation numbers must equal the charge on the molecule or ion.

(a)   −2
(b)   +4; we know that O is usually in a −2 oxidation state.
(c)   The oxidation state of an element is always zero, by definition.
(d)   −3; hydrogen is usually in a +1 oxidation state.

4.57   The sum of the oxidation numbers should be zero:

(a)   O: −2   (c)   Na: +1
       Na: +1   O: −2
       H: +1   S: +2.5
       P: +5

(b)   Ba: +2   (d)   F: −1 (Note: Except for F_2, fluorine is
       O: −2   always in a −1 oxidation state.)
       Mn: +6   Cl: +3

4.59   The sum of the oxidation numbers should be zero:

(a)   +2   (d)   +5   (g)   −1
(b)   +4   (e)   −2   (h)   −3
(c)   +3   (f)   0   (i)   −1/3

4.62   The sum of the oxidation numbers should be zero:

(a)   O: −2   (c)   O: −2
       Na: +1   Na: +1
       Cl: +1   Cl: +5

(b)   O: −2   (d)   O: −2
       Na: +1   Na: +1
       Cl: +3   Cl: +7

4.64   The sum of the oxidation numbers should be zero:

(a)   S: −2   (c)   O: −2
       Pb: +2   Sr: +2
                               I: +5

(b)   Cl: −1   (d)   S: −2
       Ti: +4   Cr: +3

4.66 (a) -2
(b) 0
(c) $+4$
(d) $+4$

4.68 (a) substance reduced (and oxidizing agent): HNO_3
substance oxidized (and reducing agent): H_3AsO_3
(b) substance reduced (and oxidizing agent): $HOCl$
substance oxidized (and reducing agent): NaI
(c) substance reduced (and oxidizing agent): $KMnO_4$
substance oxidized (and reducing agent): $H_2C_2O_4$
(d) substance reduced (and oxidizing agent): H_2SO_4
substance oxidized (and reducing agent): Al

4.71 (a) $BiO_3^- + 6H^+ + 2e^- \rightarrow Bi^{3+} + 3H_2O$ This is reduction of BiO_3^{2-}.

(b) $Pb^{2+} + 2H_2O \rightarrow PbO_2 + 4H^+ + 2e^-$ This is oxidation of Pb^{2+}.

(c) $NO_3^- + 10H^+ + 8e^- \rightarrow NH_4^+ + 3H_2O$ This constitutes reduction of NO_3^-.

(d) $6H_2O + Cl_2 \rightarrow 2ClO_3^- + 12H^+ + 10e^-$ This constitutes oxidation of Cl_2.

4.73 (a) $2S_2O_3^{2-} \rightarrow S_4O_6^{2-} + 2e^-$
$OCl^- + 2H^+ + 2e^- \rightarrow Cl^- + H_2O$
$OCl^- + 2S_2O_3^{2-} 2H^+ + \rightarrow S_4O_6^{2-} + Cl^- + H_2O$

(b) $(NO_3^- + 2H^+ + e^- \rightarrow NO_2 + H_2O) \times 2$
$Cu \rightarrow Cu^{2+} + 2e^-$
$2NO_3^- + Cu + 4H^+ \rightarrow 2NO_2 + Cu^{2+} + 2H_2O$

(c) $IO_3^- + 6H^+ + 6e^- \rightarrow I^- + 3H_2O$
$(H_2O + AsO_3^{3-} \rightarrow AsO_4^{3-} + 2H^+ + 2e^-) \times 3$
$IO_3^- + 3AsO_3^{3-} + 6H^+ + 3H_2O \rightarrow I^- + 3AsO_4^{3-} + 3H_2O + 6H^+$
which simplifies to give:
$3AsO_3^{3-} + IO_3^- \rightarrow I^- + 3AsO_4^{3-}$

(d) $SO_4^{2-} + 4H^+ + 2e^- \rightarrow SO_2 + 2H_2O$

$Zn \rightarrow Zn^{2+} + 2e^-$

$Zn + SO_4^{2-} + 4H^+ \rightarrow Zn^{2+} + SO_2 + 2H_2O$

(e) $NO_3^- + 10H^+ + 8e^- \rightarrow NH_4^+ + 3H_2O$

$(Zn \rightarrow Zn^{2+} + 2e^-) \times 4$

$NO_3^- + 4Zn + 10H^+ \rightarrow 4Zn^{2+} + NH_4^+ + 3H_2O$

(f) $2Cr^{3+} + 7H_2O \rightarrow Cr_2O_7^{2-} + 14H^+ + 6e^-$

$(BiO_3^- + 6H^+ + 2e^- \rightarrow Bi^{3+} + 3H_2O) \times 3$

$2Cr^{3+} + 3BiO_3^- + 18H^+ + 7H_2O \rightarrow Cr_2O_7^{2-} + 14H^+ + 3Bi^{3+} + 9H_2O$

which simplifies to give:

$2Cr^{3+} + 3BiO_3^- + 4H^+ \rightarrow Cr_2O_7^{2-} + 3Bi^{3+} + 2H_2O$

(g) $I_2 + 6H_2O \rightarrow 2IO_3^- + 12H^+ + 10e^-$

$(OCl^- + 2H^+ + 2e^- \rightarrow Cl^- + H_2O) \times 5$

$I_2 + 5OCl^- + H_2O \rightarrow 2IO_3^- + 5Cl^- + 2H^+$

(h) $(Mn^{2+} + 4H_2O \rightarrow MnO_4^- + 8H^+ + 5e^-) \times 2$

$(BiO_3^- + 6H^+ + 2e^- \rightarrow Bi^{3+} + 3H_2O) \times 5$

$2Mn^{2+} + 5BiO_3^- + 30H^+ + 8H_2O \rightarrow 2MnO_4^- + 5Bi^{3+} + 16H^+ + 15H_2O$

which simplifies to:

$2Mn^{2+} + 5BiO_3^- + 14H^+ \rightarrow 2MnO_4^- + 5Bi^{3+} + 7H_2O$

(i) $(H_3AsO_3 + H_2O \rightarrow H_3AsO_4 + 2H^+ + 2e^-) \times 3$

$Cr_2O_7^{2-} + 14H^+ + 6e^- \rightarrow 2Cr^{3+} + 7H_2O$

$3H_3AsO_3 + Cr_2O_7^{2-} + 3H_2O + 14H^+ \rightarrow 3H_3AsO_4 + 2Cr^{3+} + 6H^+ + 7H_2O$

which simplifies to give:

$3H_3AsO_3 + Cr_2O_7^{2-} + 8H^+ \rightarrow 3H_3AsO_4 + 2Cr^{3+} + 4H_2O$

(j) $2I^- \rightarrow I_2 + 2e^-$

$HSO_4^- + 3H^+ + 2e^- \rightarrow SO_2 + 2H_2O$

$2I^- + HSO_4^- + 3H^+ \rightarrow I_2 + SO_2 + 2H_2O$

4.75   For redox reactions in basic solution, we proceed to balance the half reactions as if they were in acid solution, and then add enough OH^- to each side of the resulting equation in order to neutralize (titrate) all of the H^+. This gives a corresponding amount of water $(H^+ + OH^- \rightarrow H_2O)$ on one side of the equation, and an excess of OH^- on the other side of the equation, as befits a reaction in basic solution.

(a) $(CrO_4^{2-} + 4H^+ + 3e^- \rightarrow CrO_2^- + 2H_2O) \times 2$

$(S^{2-} \rightarrow S + 2e^-) \times 3$

$2CrO_4^{2-} + 3S^{2-} + 8H^+ \rightarrow 2CrO_2^- + 2S + 4H_2O$

Adding $8OH^-$ to both sides of the above equation we obtain:

$2CrO_4^{2-} + 3S^{2-} + 8H_2O \rightarrow 2CrO_2^- + 8OH^- + 3S + 4H_2O$

which simplifies to:

$2CrO_4^{2-} + 3S^{2-} + 4H_2O \rightarrow 2CrO_2^- + 3S + 8OH^-$

(b) $(C_2O_4^{2-} \rightarrow 2CO_2 + 2e^-) \times 3$

$(MnO_4^- + 4H^+ + 3e^- \rightarrow MnO_2 + 2H_2O) \times 2$

$3C_2O_4^{2-} + 2MnO_4^- + 8H^+ \rightarrow 6CO_2 + 2MnO_2 + 4H_2O$

Adding $8OH^-$ to both sides of the above equation we get:

$3C_2O_4^{2-} + 2MnO_4^- + 8H_2O \rightarrow 6CO_2 + 2MnO_2 + 4H_2O + 8OH^-$

which simplifies to give:

$3C_2O_4^{2-} + 2MnO_4^- + 4H_2O \rightarrow 6CO_2 + 2MnO_2 + 8OH^-$

(c) $(ClO_3^- + 6H^+ + 6e^- \rightarrow Cl^- + 3H_2O) \times 4$

$(N_2H_4 + 2H_2O \rightarrow 2NO + 8H^+ + 8e^-) \times 3$

$4ClO_3^- + 3N_2H_4 + 24H^+ + 6H_2O \rightarrow 4Cl^- + 6NO + 12H_2O + 24H^+$

which needs no OH^-, because it simplifies directly to:

$4ClO_3^- + 3N_2H_4 \rightarrow 4Cl^- + 6NO + 6H_2O$

(d) $\quad NiO_2 + 2H^+ + 2e^- \rightarrow Ni(OH)_2$

$\quad\quad 2Mn(OH)_2 \rightarrow Mn_2O_3 + H_2O + 2H^+ + 2e^-$

$\quad\quad NiO_2 + 2Mn(OH)_2 \rightarrow Ni(OH)_2 + Mn_2O_3 + H_2O$

(e) $\quad (SO_3^{2-} + H_2O \rightarrow SO_4^{2-} + 2H^+ + 2e^-) \times 3$

$\quad\quad (MnO_4^- + 4H^+ + 3e^- \rightarrow MnO_2 + 2H_2O) \times 2$

$\quad\quad 3SO_3^{2-} + 3H_2O + 8H^+ + 2MnO_4^- \rightarrow 3SO_4^{2-} + 6H^+ + 2MnO_2 + 4H_2O$

Adding $8OH^-$ to both sides of the equation we obtain:

$\quad\quad 3SO_3^{2-} + 11H_2O + 2MnO_4^- \rightarrow 3SO_4^{2-} + 10H_2O + 2MnO_2 + 2OH^-$

which simplifies to:

$\quad\quad 3SO_3^{2-} + 2MnO_4^- + H_2O \rightarrow 3SO_4^{2-} + 2MnO_2 + 2OH^-$

(f) $\quad (CrO_2^- + 2H_2O \rightarrow CrO_4^{2-} + 4H^+ + 3e^-) \times 2$

$\quad\quad (S_2O_8^{2-} + 2e^- \rightarrow 2SO_4^{2-}) \times 3$

$\quad\quad 3S_2O_8^{2-} + 2CrO_2^- + 4H_2O \rightarrow 2CrO_4^{2-} + 6SO_4^{2-} + 8H^+$

Adding $8OH^-$ to both sides of this equation:

$\quad\quad 3S_2O_8^{2-} + 2CrO_2^- + 4H_2O + 8OH^- \rightarrow 2CrO_4^{2-} + 6SO_4^{2-} + 8H_2O$

which simplifies to give:

$\quad\quad 3S_2O_8^{2-} + 2CrO_2^- + 8OH^- \rightarrow 2CrO_4^{2-} + 6SO_4^{2-} + 4H_2O$

(g) $\quad (SO_3^{2-} + H_2O \rightarrow SO_4^{2-} + 2H^+ + 2e^-) \times 3$

$\quad\quad (CrO_4^{2-} + 4H^+ + 3e^- \rightarrow CrO_2^- + 2H_2O) \times 2$

$\quad\quad 3SO_3^{2-} + 2CrO_4^{2-} + 8H^+ + 3H_2O \rightarrow 3SO_4^{2-} + 2CrO_2^- + 6H^+ + 4H_2O$

Adding $8OH^-$ to both sides of the equation we get:

$\quad\quad 3SO_3^{2-} + 2CrO_4^{2-} + 11H_2O \rightarrow 3SO_4^{2-} + 2CrO_2^- + 2OH^- + 10H_2O$

which simplifies to:

$\quad\quad 3SO_3^{2-} + 2CrO_4^{2-} + H_2O \rightarrow 3SO_4^{2-} + 2CrO_2^- + 2OH^-$

(h) $\quad (O_2 + 2H^+ + 2e^- \rightarrow H_2O_2) \times 2$

$\quad\quad N_2H_4 \rightarrow N_2 + 4H^+ + 4e^-$

$\quad\quad 2O_2 + 4H^+ + N_2H_4 \rightarrow 2H_2O_2 + N_2 + 4H^+$

which simplifies to:

$$2O_2 + N_2H_4 \rightarrow 2H_2O_2 + N_2$$

(i) $(Fe(OH)_2 + OH^- \rightarrow Fe(OH)_3 + e^-) \times 4$

$$O_2 + 2H_2O + 4e^- \rightarrow 4OH^-$$

$$4Fe(OH)_2 + O_2 + 4OH^- + 2H_2O \rightarrow 4Fe(OH)_3 + 4OH^-$$

which simplifies to:

$$4Fe(OH)_2 + O_2 + 2H_2O \rightarrow 4Fe(OH)_3$$

(j) $(Au + 4CN^- \rightarrow Au(CN)_4^- + 3e^-) \times 4$

$$(O_2 + 2H_2O + 4e^- \rightarrow 4OH^-) \times 3$$

$$4Au + 16CN^- + 3O_2 + 6H_2O \rightarrow 4Au(CN)_4^- + 12OH^-$$

4.77 $(OCl^- + 2H^+ + 2e^- \rightarrow Cl^- + H_2O) \times 4$

$$S_2O_3^{2-} + 5H_2O \rightarrow 2SO_4^{2-} + 10H^+ + 8e^-$$

$$4OCl^- + S_2O_3^{2-} + 5H_2O + 8H^+ \rightarrow 4Cl^- + 2SO_4^{2-} + 10H^+ + 4H_2O$$

which simplifies to:

$$4OCl^- + S_2O_3^{2-} + H_2O \rightarrow 4Cl^- + 2SO_4^{2-} + 2H^+$$

4.81 For each of the following recall that molarity is defines as moles of solute divided by liters of solution.

(a)
$$M\ NaOH = \left(\frac{4.00\ g\ NaOH}{100.0\ mL\ NaOH\ soln}\right)\left(\frac{1\ mol\ NaOH}{40.00\ g\ NaOH}\right)$$
$$\times \left(\frac{1000\ mL\ NaOH\ soln}{1\ L\ NaOH\ soln}\right)$$
$$= 1.00\ M\ NaOH$$

(b)
$$M\ CaCl_2 = \left(\frac{16.0\ g\ CaCl_2}{250.0\ mL\ CaCl_2\ soln}\right)\left(\frac{1\ mol\ CaCl_2}{110.98\ g\ CaCl_2}\right)$$
$$\times \left(\frac{1000\ mL\ CaCl_2\ soln}{1\ L\ CaCl_2\ soln}\right)$$
$$= 0.577\ M\ CaCl_2$$

(c)

$$M \ KOH = \left(\frac{14.0 \ g \ KOH}{75.0 \ mL \ KOH \ soln} \right) \left(\frac{1 \ mol \ KOH}{56.11 \ g \ KOH} \right)$$

$$\times \left(\frac{1000 \ mL \ KOH \ soln}{1 \ L \ KOH \ soln} \right)$$

$$= 3.33 \ M \ KOH$$

(d)

$$M \ H_2C_2O_4 = \left(\frac{6.75 \ g \ H_2C_2O_4}{500 \ mL \ H_2C_2O_4 \ soln} \right) \left(\frac{1 \ mol \ H_2C_2O_4}{90.0 \ g \ H_2C_2O_4} \right)$$

$$\times \left(\frac{1000 \ mL \ H_2C_2O_4 \ soln}{1 \ L \ H_2C_2O_4 \ soln} \right)$$

$$= 0.2 \ M \ H_2C_2O_4$$

(Note: If we assume 3 sig figs we get $0.150 \ M \ H_2C_2O_4$)

4.83 (a)

$$\# \ g \ NaCl = (125 \ mL \ NaCl \ soln) \left(\frac{1 \ L \ NaCl \ soln}{1000 \ mL \ NaCl \ soln} \right)$$

$$\times \left(\frac{0.200 \ mol \ NaCl}{1 \ L \ NaCl \ soln} \right) \left(\frac{58.44 \ g \ NaCl}{1 \ mol \ NaCl} \right)$$

$$= 1.46 \ g \ NaCl$$

(b)

$$\# \ g \ C_6H_{12}O_6 = (250 \ mL \ C_6H_{12}O_6 \ soln) \left(\frac{1 \ L \ C_6H_{12}O_6 \ soln}{1000 \ mL \ C_6H_{12}O_6 \ soln} \right)$$

$$\times \left(\frac{0.360 \ mol \ C_6H_{12}O_6}{1 \ L \ C_6H_{12}O_6 \ soln} \right) \left(\frac{180.2 \ g \ C_6H_{12}O_6}{1 \ mol \ C_6H_{12}O_6} \right)$$

$$= 16 \ g \ C_6H_{12}O_6$$

(c)

$$\# \, g \, H_2SO_4 = (250 \, mL \, H_2SO_4 \, soln)\left(\frac{1 \, L \, H_2SO_4 \, soln}{1000 \, mL \, H_2SO_4 \, soln}\right)$$

$$\times \left(\frac{0.250 \, mol \, H_2SO_4}{1 \, L \, H_2SO_4 \, soln}\right)\left(\frac{98.08 \, g \, H_2SO_4}{1 \, mol \, H_2SO_4}\right)$$

$$= 6.1 \, g \, H_2SO_4$$

4.85 (a) $KOH \rightarrow K^+ + OH^-$

1.25 mol/L × 0.0350 L = 0.0438 mol KOH

0.0438 mol KOH × 1 mol OH^-/mol KOH = 0.0438 mol OH^-

0.0438 mol KOH × 1 mol K^+/mol KOH = 0.0438 mol K^+

(b) $CaCl_2 \rightarrow Ca^{2+} + 2Cl^-$

0.45 mol/L × 0.0323 L = 0.015 mol $CaCl_2$

0.015 mol $CaCl_2$ × 1 mol Ca^{2+}/mol $CaCl_2$ = 0.015 mol Ca^{2+}

0.015 mol $CaCl_2$ × 2 mol Cl^-/mol $CaCl_2$ = 0.030 mol Cl^-

(c) $(NH_4)_2CO_3 \rightarrow 2NH_4^+ + CO_3^{2-}$

0.40 mol/L × 0.0185 L = 0.074 mol $(NH_4)_2CO_3$

0.074 mol $(NH_4)_2CO_3$ × 2 mol NH_4^+/mol $(NH_4)_2CO_3$ = 0.15 mol NH_4^+

0.074 mol $(NH_4)_2CO_3$ × 1 mol CO_3^{2-}/mol $(NH_4)_2CO_3$ = 0.074 mol CO_3^{2-}

(d) $Al_2(SO_4)_3 \rightarrow 2Al^{3+} + 3SO_4^{2-}$

0.35 mol/L × 0.0300 L = 0.011 mol $Al_2(SO_4)_3$

0.011 mol $Al_2(SO_4)_3$ × 2 mol Al^{3+}/mol $Al_2(SO_4)_3$ = 0.022 mol Al^{3+}

0.012 mol $Al_2(SO_4)_3$ × 3 mol SO_4^{2-}/mol $Al_2(SO_4)_3$ = 0.033 mol SO_4^{2-}

4.86   (a) $Cr(NO_3)_2 \rightarrow Cr^{2+} + 2NO_3^{-}$

$$M\ Cr^{2+} = \left(\frac{0.25\ mol\ Cr(NO_3)_2}{1\ L\ Cr(NO_3)_2\ soln}\right)\left(\frac{1\ mol\ Cr^{2+}}{1\ mol\ Cr(NO_3)_2}\right) = 0.25\ M\ Cr^{2+}$$

$$M\ NO_3^{-} = \left(\frac{0.25\ mol\ Cr(NO_3)_2}{1\ L\ Cr(NO_3)_2\ soln}\right)\left(\frac{2\ mol\ NO_3^{-}}{1\ mol\ Cr(NO_3)_2}\right) = 0.50\ M\ NO_3^{-}$$

(b) $CuSO_4 \rightarrow Cu^{2+} + SO_4^{2-}$

$$M\ Cu^{2+} = \left(\frac{0.10\ mol\ CuSO_4}{1\ L\ CuSO_4\ soln}\right)\left(\frac{1\ mol\ Cu^{2+}}{1\ mol\ CuSO_4}\right) = 0.10\ M\ Cu^{2+}$$

$$M\ SO_4^{2-} = \left(\frac{0.10\ mol\ CuSO_4}{1\ L\ CuSO_4\ soln}\right)\left(\frac{1\ mol\ SO_4^{2-}}{1\ mol\ CuSO_4}\right) = 0.10\ M\ SO_4^{2-}$$

(c) $Na_3PO_4 \rightarrow 3Na^{+} + PO_4^{3-}$

$$M\ Na^{+} = \left(\frac{0.16\ mol\ Na_3PO_4}{1\ L\ Na_3PO_4\ soln}\right)\left(\frac{3\ mol\ Na^{+}}{1\ mol\ Na_3PO_4}\right) = 0.48\ M\ Na^{+}$$

$$M\ PO_4^{3-} = \left(\frac{0.16\ mol\ Na_3PO_4}{1\ L\ Na_3PO_4\ soln}\right)\left(\frac{1\ mol\ PO_4^{3-}}{1\ mol\ Na_3PO_4}\right) = 0.16\ M\ PO_4^{3-}$$

(d) $Al_2(SO_4)_3 \rightarrow 2Al^{3+} + 3SO_4^{2-}$

$$M\ Al^{3+} = \left(\frac{0.075\ mol\ Al_2(SO_4)_3}{1\ L\ Al_2(SO_4)_3\ soln}\right)\left(\frac{2\ mol\ Al^{3+}}{1\ mol\ Al_2(SO_4)_3}\right) = 0.15\ M\ Al^{3+}$$

$$M\ SO_4^{2-} = \left(\frac{0.075\ mol\ Al_2(SO_4)_3}{1\ L\ Al_2(SO_4)_3\ soln}\right)\left(\frac{3\ mol\ SO_4^{2-}}{1\ mol\ Al_2(SO_4)_3}\right) = 0.23\ M\ SO_4^{2-}$$

4.88 $M\ Na_3PO_4 = \left(\frac{0.21\ mol\ Na^{+}}{1\ L\ Na_3PO_4\ soln}\right)\left(\frac{1\ mol\ Na_3PO_4}{3\ mol\ Na^{+}}\right) = 0.070\ M\ Na_3PO_4$

4.90   Use equation 4.3

$$M_{dil} = \frac{V_{concd} \bullet M_{concd}}{V_{dil}} = \frac{(25.0\ mL)(0.56\ M\ H_2SO_4)}{125\ mL} = 0.11\ M\ H_2SO_4$$

4.92 Use equation 4.3

$$V_{dil} = \frac{V_{concd} \cdot M_{concd}}{M_{dil}} = \frac{(25.0 \text{ mL})(18.0 \text{ M H}_2\text{SO}_4)}{1.50 \text{ M H}_2\text{SO}_4} = 3.00 \times 10^2 \text{ mL}$$

4.94 In order to solve this problem we must recall two things; molarity is defined as the number of moles of solute per liter of solution and that molarity times volume equals the number of moles. Set up the following algebraic equation:

$$\frac{(50.0 \text{ mL})(0.40 \text{ M}) + (x \text{ mL})(0.10 \text{ M})}{(50.0 \text{ mL} + x \text{ mL})} = 0.25 \text{ M}$$

The numerator of this expression is the total number of moles of solution and the denominator is the total volume of the solution. We can now solve this equation for x and we find that $x = 5.0 \times 10^1$ mL.

4.96

$$M \text{ H}_2\text{SO}_4 =$$

$$\left[\frac{(26.04 \text{ mL NaOH soln})\left(\frac{1 \text{ L NaOH soln}}{1000 \text{ mL NaOH soln}}\right)\left(\frac{0.1024 \text{ mol NaOH}}{1 \text{ L NaOH soln}}\right)\left(\frac{1 \text{ mol H}_2\text{SO}_4}{2 \text{ mol NaOH}}\right)}{(12.88 \text{ mL H}_2\text{SO}_4 \text{ soln})\left(\frac{1 \text{ L H}_2\text{SO}_4 \text{ soln}}{1000 \text{ mL H}_2\text{SO}_4 \text{ soln}}\right)} \right]$$

$$= 0.1035 \text{ M H}_2\text{SO}_4$$

4.98 First, determine the number of moles of $BaCl_2$ that are to react:

$0.0500 \text{ L} \times 0.125 \text{ mol/L} = 0.00625 \text{ mol BaCl}_2$

Next, determine the number of moles of Epsom salts that are required, based on the balanced equation:

$MgSO_4 \cdot 7H_2O + BaCl_2 \rightarrow BaSO_4 + MgCl_2 + 7H_2O$

$$\# \text{ mol MgSO}_4 \cdot 7H_2O = (6.25 \times 10^{-3} \text{ mol BaCl}_2)\left(\frac{1 \text{ mol MgSO}_4 \cdot 7H_2O}{1 \text{ mol BaCl}_2}\right)$$

$$= 6.25 \times 10^{-3} \text{ mol MgSO}_4 \cdot 7H_2O$$

Finally, calculate the required mass:

$0.00625 \text{ mol MgSO}_4 \cdot 7H_2O \times 246.5 \text{ g/mol} = 1.54 \text{ g MgSO}_4 \cdot 7H_2O$

4.100 $NaHCO_3 + HCl \rightarrow NaCl + H_2O + CO_2$

$$\# \text{ g } NaHCO_3 = (162 \text{ ml HCl soln})\left(\frac{0.052 \text{ mol HCl}}{1000 \text{ mL HCl soln}}\right)$$

$$\times \left(\frac{1 \text{ mol } NaHCO_3}{1 \text{ mol HCl}}\right)\left(\frac{84.01 \text{ g } NaHCO_3}{1 \text{ mol } NaHCO_3}\right)$$

$$= 0.71 \text{ g } NaHCO_3$$

4.102

$$\# \text{ mL } (NH_4)_2 SO_4 \text{ soln } =$$

$$(50.0 \text{ ml NaOH soln})\left(\frac{1.00 \text{ mol NaOH}}{1000 \text{ mL NaOH soln}}\right)\left(\frac{1 \text{ mol } OH^-}{1 \text{ mol NaOH}}\right)$$

$$\times \left(\frac{1 \text{ mol } NH_4^+}{1 \text{ mol } OH^-}\right)\left(\frac{1 \text{ mol } (NH_4)_2 SO_4}{2 \text{ mol } NH_4^+}\right)\left(\frac{1000 \text{ mL } (NH_4)_2 SO_4 \text{ soln}}{0.250 \text{ mol } (NH_4)_2 SO_4}\right)$$

$$= 1.00 \times 10^2 \text{ mL } (NH_4)_2 SO_4 \text{ soln}$$

4.104 The equation for the reaction indicates that the two materials react in equimolar amounts, i.e. the stoichiometry is 1 to 1:

$$AgNO_3(aq) + NaCl(aq) \rightarrow AgCl(s) + NaNO_3(aq)$$

(a) Because this reaction is 1:1, we can see by inspection that the $AgNO_3$ is the limiting reagent. We know this because the concentration of the $AgNO_3$ is lower than the NaCl. Since we start with equal volumes, there are fewer moles of the $AgNO_3$.

$$\# \text{ mol AgCl } = (25.0 \text{ mL } AgNO_3 \text{ soln})\left(\frac{0.320 \text{ mol } AgNO_3}{1000 \text{ mL } AgNO_3 \text{ soln}}\right)$$

$$\times \left(\frac{1 \text{ mol AgCl}}{1 \text{ mol } AgNO_3}\right)$$

$$= 8.00 \times 10^{-3} \text{ mol AgCl}$$

(b) Assuming that AgCl is essentially insoluble, the concentration of silver ion can be said to be zero since all of the $AgNO_3$ reacted. The number of moles of chloride ion would be reduced by the precipitation of 8.00×10^{-3} mol AgCl, such that the final number of moles of chloride ion would be:

$$0.0250 \text{ L} \times 0.440 \text{ mol/L} - 8.00 \times 10^{-3} \text{ mol} = 3.0 \times 10^{-3} \text{ mol } Cl^-$$

The final concentration of Cl^- is, therefore:

$$3.0 \times 10^{-3} \text{ mol} \div 0.0500 \text{ L} = 0.060 \text{ M } Cl^-$$

All of the original number of moles of NO_3^- and of Na^+ would still be present in solution, and their concentrations would be:

For NO_3^-:

$$\# \text{ M } NO_3^- = \frac{(25.0 \text{ mL AgNO}_3 \text{ soln})\left(\dfrac{0.320 \text{ mol AgNO}_3}{1000 \text{ mL AgNO}_3 \text{ soln}}\right)\left(\dfrac{1 \text{ mol } NO_3^-}{1 \text{ mol AgNO}_3}\right)}{(50.0 \text{ mL soln})\left(\dfrac{1 \text{ L soln}}{1000 \text{ mL soln}}\right)}$$

$$= 0.160 \text{ M } NO_3^-$$

For Na^+:

$$\# \text{ M } Na^+ = \frac{(25.0 \text{ mL NaCl soln})\left(\dfrac{0.440 \text{ mol NaCl}}{1000 \text{ mL NaCl soln}}\right)\left(\dfrac{1 \text{ mol } Na^+}{1 \text{ mol NaCl}}\right)}{(50.0 \text{ mL soln})\left(\dfrac{1 \text{ L soln}}{1000 \text{ mL soln}}\right)}$$

$$= 0.220 \text{ M } Na^+$$

4.106

$$\# \text{ mL KMnO}_4 = (40.0 \text{ mL SnCl}_2)\left(\frac{0.250 \text{ mol SnCl}_2}{1000 \text{ mL SnCl}_2}\right)\left(\frac{1 \text{ mol Sn}^{2+}}{1 \text{ mol SnCl2}}\right)$$

$$\times \left(\frac{2 \text{ mol MnO}_4^-}{5 \text{ mol Sn}^{2+}}\right)\left(\frac{1 \text{ mol KMnO}_4}{1 \text{ mol MnO}_4^-}\right)\left(\frac{1000 \text{ mL KMnO}_4}{0.230 \text{ mol KMnO}_4}\right)$$

$$= 17.4 \text{ mL KMnO}_4$$

4.108

$$\# \text{ g PbO}_2 = (15.0 \text{ g Cl}_2)\left(\frac{1 \text{ mol Cl}_2}{70.91 \text{ g Cl}_2}\right)\left(\frac{1 \text{ mol PbO}_2}{1 \text{ mol Cl}_2}\right)\left(\frac{239.2 \text{ g PbO}_2}{1 \text{ mol PbO}_2}\right)$$

$$= 50.6 \text{ g PbO}_2$$

4.110 (a) $[Mn^{2+}(aq) + 4H_2O(\ell) \rightarrow MnO_4^-(aq) + 8H^+(aq) + 5e^-] \times 2$

$[BiO_3^-(aq) + 6H^+(aq) + 2e^- \rightarrow Bi^{3+}(aq) + 3H_2O(\ell)] \times 5$

$2Mn^{2+}(aq) + 5BiO_3^-(aq) + 8H_2O(\ell) + 30H^+(aq) \rightarrow$

$2MnO_4^-(aq) + 5Bi^{3+}(aq) + 15H_2O(\ell) + 16H^+(aq)$

which simplifies to give:

$2Mn^{2+}(aq) + 5BiO_3^-(aq) + 14H^+(aq) \rightarrow 2MnO_4^-(aq) + 5Bi^{3+}(aq) + 7H_2O(\ell)$

(b)

$$\# \text{ g NaBiO}_3 = \left(18.5 \text{ g Mn(NO}_3)_2\right)\left(\frac{1 \text{ mol Mn(NO}_3)_2}{179.0 \text{ g Mn(NO}_3)_2}\right)\left(\frac{1 \text{ mol Mn}^{2+}}{1 \text{ mol Mn(NO}_3)_2}\right)$$

$$\times \left(\frac{5 \text{ mol BiO}_3^-}{2 \text{ mol Mn}^{2+}}\right)\left(\frac{1 \text{ mol NaBiO}_3}{1 \text{ mol BiO}_3^-}\right)\left(\frac{280.0 \text{ g NaBiO}_3}{1 \text{ mol NaBiO}_3}\right)$$

$$= 72.3 \text{ g NaBiO}_3$$

4.113 First, calculate the number of moles HCl based on the titration according to the following equation:

$$NaOH(aq) + HCl(aq) \rightarrow NaCl(aq) + H_2O(\ell)$$

$$\# \text{ mol HCl} = (23.25 \text{ mL NaOH})\left(\frac{0.105 \text{ mol NaOH}}{1000 \text{ mL NaOH}}\right)\left(\frac{1 \text{ mol HCl}}{1 \text{ mol NaOH}}\right)$$

$$= 2.44 \times 10^{-3} \text{ mol HCl}$$

Next, determine the concentration of the HCl solution:

$2.44 \times 10^{-3} \text{ mol} \div 0.02145 \text{ L} = 0.114 \text{ M HCl}$

4.115 Since lactic acid is monoprotic, it reacts with sodium hydroxide on a one to one mole basis:

$$\# \text{ mol HC}_3\text{H}_5\text{O}_3 = (17.25 \text{ mL NaOH})\left(\frac{0.155 \text{ mol NaOH}}{1000 \text{ mL NaOH}}\right)\left(\frac{1 \text{ mol HC}_3\text{H}_5\text{O}_3}{1 \text{ mol NaOH}}\right)$$

$$= 2.67 \times 10^{-3} \text{ mol HC}_3\text{H}_5\text{O}_3$$

4.117 (a) From the Law of Conservation of Mass we know that the amount of Cl we ended with must equal the amount initially:

$$\# \text{ g Cl} = (0.678 \text{ g AgCl})\left(\frac{1 \text{ mol AgCl}}{143.3 \text{ g AgCl}}\right)\left(\frac{1 \text{ mol Cl}}{1 \text{ mol AgCl}}\right)\left(\frac{35.45 \text{ g Cl}}{1 \text{ mol Cl}}\right)$$
$$= 0.168 \text{ g Cl}$$

(b) Determine moles and then relative moles of Fe and Cl:

$0.168 \text{ g Cl} \div 35.45 \text{ g/mol} = 4.74 \times 10^{-3} \text{ mol Cl}$

$(0.300 - 0.168) \text{ g Fe} \div 55.85 \text{ g/mol Fe} = 2.36 \times 10^{-3} \text{ mol Fe}$

For Cl, the relative number of moles is:

$4.74 \times 10^{-3} \text{ mol}/2.36 \times 10^{-3} \text{ mol} = 2.01$

For Fe, the relative number of moles is:

$2.36 \times 10^{-3} \text{ mol}/2.36 \times 10^{-3} \text{ mol} = 1.00$

and the formula is seen to be $FeCl_2$.

4.119 $0.02940 \text{ L KOH} \times 0.0300 \text{ mol/L} = 8.82 \times 10^{-4} \text{ mol KOH}$

$8.82 \times 10^{-4} \text{ mol KOH} \times (1 \text{ mol aspirin}/1 \text{ mol KOH}) = 8.82 \times 10^{-4} \text{ mol aspirin}$

$8.82 \times 10^{-4} \text{ mol aspirin} \times 180.2 \text{ g/mol} = 0.159 \text{ g aspirin}$

$0.159 \text{ g aspirin}/0.250 \text{ g total} \times 100 = 63.6 \text{ \% aspirin}$

4.121 (a) $2CrO_4^{2-} + 3SO_3^{2-} + H_2O \rightarrow 2CrO_2^{-} + 3SO_4^{2-} + 2OH^{-}$

(b)

$$\# \text{ mol } CrO_4^{2-} = (3.18 \text{ g Na}_2SO_3)\left(\frac{1 \text{ mol Na}_2SO_3}{126.04 \text{ g Na}_2SO_3}\right)$$
$$\times \left(\frac{1 \text{ mol SO}_3^{2-}}{1 \text{ mol Na}_2SO_3}\right)\left(\frac{2 \text{ mol CrO}_4^{2-}}{3 \text{ mol SO}_3^{2-}}\right)$$
$$= 1.68 \times 10^{-2} \text{ mol } CrO_4^{2-}$$

(c) Since there is one mole of Cr in each mole of CrO_4^{2-}, then the above number of moles of CrO_4^{2-} is also equal to the number of moles of Cr that were present:

$0.0168 \text{ mol Cr} \times 52.00 \text{ g/mol} = 0.875 \text{ g Cr in the original alloy.}$

(d) $(0.875 \text{ g}/3.450 \text{ g}) \times 100 = 25.4 \text{ \% Cr}$

4.123

$$\text{\# g NaNO}_2 \ = \ (12.15 \text{ mL KMnO}_4)\left(\frac{0.01000 \text{ mol KMnO}_4}{1000 \text{ mL KMnO}_4}\right)\left(\frac{1 \text{ mol MnO}_4{}^{2-}}{1 \text{ mol KMnO}_4}\right)$$

$$\times \ \left(\frac{5 \text{ mol HNO}_2}{2 \text{ mol MnO}_4{}^{2-}}\right)\left(\frac{1 \text{ mol NaNO}_2}{1 \text{ mol HNO}_2}\right)\left(\frac{68.995 \text{ g NaNO}_2}{1 \text{ mol NaNO}_2}\right)$$

$$= \ 2.096 \text{ X } 10^{-2} \text{ g NaNO}_2$$

$$\% \text{ NaNO}_2 = (2.096 \text{ X } 10^{-2} \text{ g NaNO}_2 / 1.000 \text{ g sample}) \times 100 = 2.096 \ \%$$

4.125 **(a)**

$$\text{\# mol I}_3{}^- \ = \ (29.25 \text{ mL Na}_2\text{S}_2\text{O}_3)\left(\frac{0.3000 \text{ mol Na}_2\text{S}_2\text{O}_3}{1000 \text{ mL Na}_2\text{S}_2\text{O}_3}\right)\left(\frac{1 \text{ mol I}_3{}^-}{2 \text{ mol Na}_2\text{S}_2\text{O}_3}\right)$$

$$= \ 4.388 \text{ X } 10^{-3} \text{ mol I}_3{}^-$$

(b)

$$\text{\# mol NO}_2{}^- \ = \ (4.388 \text{ X } 10^{-3} \text{ mol I}_3{}^-)\left(\frac{2 \text{ mol NO}_2{}^-}{1 \text{ mol I}_3{}^-}\right)$$

$$= \ 8.776 \text{ X } 10^{-3} \text{ mol NO}_2{}^-$$

(c)

$$\text{\# g NaNO}_2 \ = \ (8.776 \text{ X } 10^{-3} \text{ mol NO}_2{}^-)\left(\frac{1 \text{ mol NaNO}_2}{1 \text{ mol NO}_2{}^-}\right)\left(\frac{68.995 \text{ g NaNO}_2}{1 \text{ mol NaNO}_2}\right)$$

$$= \ 0.6055 \text{ g NaNO}_2$$

$$\% \text{ NaNO}_2 = (0.6055 \text{ g NaNO}_2 / 1.104 \text{ g sample}) \times 100 = 54.85 \ \% \text{ NaNO}_2$$

4.127 **(a)**

$$\text{\# mol Cu}^{2+} \ = \ (29.96 \text{ mL S}_2\text{O}_3{}^{2-})\left(\frac{0.02100 \text{ mol S}_2\text{O}_3{}^{2-}}{1000 \text{ mL S}_2\text{O}_3{}^{2-}}\right)$$

$$\times \ \left(\frac{1 \text{ mol I}_3{}^-}{2 \text{ mol S}_2\text{O}_3{}^{2-}}\right)\left(\frac{2 \text{ mol Cu}^{2+}}{1 \text{ mol I}_3{}^-}\right)$$

$$= \ 6.292 \text{ X } 10^{-4} \text{ mol Cu}^{2+}$$

$$\text{\# g Cu} = (6.292 \times 10^{-4} \text{ mol Cu}) \times (63.546 \text{ g Cu/mol Cu})$$
$$= 3.998 \times 10^{-2} \text{ g Cu}$$

$$\% \text{ Cu} = (3.998 \times 10^{-2} \text{ g Cu}/0.4225 \text{ g sample}) \times 100 = 9.463 \%$$

(b)

$$\text{\# g CuCO}_3 = (6.292 \times 10-4 \text{ mol CuCO}_3)\left(\frac{123.56 \text{ g CuCO}_3}{1 \text{ mol CuCO}_3}\right)$$

$$= 0.07774 \text{ g CuCO}_3$$

$$\% \text{ CuCO}_3 = \left(\frac{0.07774 \text{ g CuCO}_3}{0.4225 \text{ g sample}}\right) \times 100 = 18.40 \%$$

4.128 (a) strong electrolyte   (e) weak electrolyte
(b) nonelectrolyte   (f) nonelectrolyte
(c) strong electrolyte   (g) strong electrolyte
(d) non electrolyte   (h) weak electrolyte

4.130 (a) molecular: $Na_2S(aq) + H_2SO_4(aq) \rightarrow H_2S(g) + Na_2SO_4(aq)$

net ionic: $S^{2-}(aq) + 2H^+(aq) \rightarrow H_2S(g)$

(b) molecular: $LiHCO_3(aq) + HNO_3(aq) \rightarrow LiNO_3(aq) + H_2O(\ell) + CO_2(g)$

net ionic: $H^+(aq) + HCO_3^-(aq) \rightarrow H_2O(\ell) + CO_2(g)$

(c) molecular: $(NH_4)_3PO_4(aq) + 3KOH(aq) \rightarrow$
$$K_3PO_4(aq) + 3NH_3(aq) + 3H_2O(\ell)$$

net ionic: $NH_4^+(aq) + OH^-(aq) \rightarrow NH_3(aq) + H_2O(\ell)$

(d) molecular: $K_2SO_3(aq) + HCl(aq) \rightarrow KHSO_3(aq) + KCl(aq)$

net ionic: $SO_3^{2-}(aq) + H^+(aq) \rightarrow HSO_3^-(aq)$

(e) molecular: $BaCO_3(s) + 2HBr(aq) \rightarrow BaBr_2(aq) + CO_2(g) + H_2O(\ell)$

net ionic: $BaCO_3(s) + 2H^+(aq) \rightarrow Ba^{2+}(aq) + CO_2(g) + H_2O(\ell)$

(f) no reaction

4.132 Start with a balanced net ionic equation:

$$CO_3^{2-}(aq) + 2H^+(aq) \rightarrow H_2O(\ell) + CO_2(g)$$

$$\# \text{ g Na}_2CO_3 = (43.6 \text{ mL HCl})\left(\frac{0.116 \text{ mol HCl}}{1000 \text{ mL HCl}}\right)\left(\frac{1 \text{ mol Na}_2CO_3}{2 \text{ mol HCl}}\right)\left(\frac{105.99 \text{ g Na}_2CO_3}{1 \text{ mol Na}_2CO_3}\right)$$

$$= 0.268 \text{ g Na}_2CO_3$$

% Na$_2$CO$_3$ = (0.268 g Na$_2$CO$_3$ / 0.326 g sample) × 100 = 82.2%

4.134 The oxidation state of cerium in the reactant ion is +4. The concentration of this ion in the reactant solution is:

0.0150 M × 0.02500 L = 3.75 × 10^{-4} mol Ce^{IV}

The number of moles of electrons that come from the Fe^{2+} reducing agent is:

0.0320 M × 0.02344 L = 7.50 × 10^{-4} mol e$^-$

The ratio of moles of electrons to moles of Ce^{IV} reactant is therefore 2:1, and we conclude that the product is Ce^{II} (Ce^{2+}).

CHAPTER FIVE
Review Exercises

5.12 $\# \text{ kJ} = \left(175 \text{ g H}_2\text{O}\right)\left(4.1796 \dfrac{J}{g \, °C}\right)\left(25.0 \, °C - 15.0 \, °C\right)\left(\dfrac{1 \text{ kJ}}{1000 \text{ J}}\right) = 7.31 \text{ kJ}$

5.14 $\# \text{ J} = 0.4498 \text{ J g}^{-1} \, °C^{-1} \times 15.0 \text{ g} \times 20.0 \, °C = 135 \text{ J}$

5.16 $\text{X J g}^{-1} \, °C^{-1} \times 20.0 \text{ g} \times 10.0 \, °C = 47.0 \text{ J}$
        $\text{X} = 0.235 \text{ J g}^{-1} \, °C^{-1}$, which is the value in Table 5.1 for silver.

5.17 $2.0 \text{ J g}^{-1} \, °C^{-1} \times 500 \text{ mL} \times 0.91 \text{ g ml}^{-1} \times \Delta T = 1.25 \text{ X } 10^3 \text{ J}$
        $\Delta T = 1.4 \, °C$
        $t_{final} = 25.0 + 1.4 = 26.4 \, °C$

5.19   (a) $4.1796 \text{ J g}^{-1} \, °C^{-1} \times 200 \text{ g} \times 1.50 \, °C = 1.25 \text{ X } 10^3 \text{ J}$
        (b) $1.25 \text{ X } 10^3 \text{ J}$
        (c) $1.25 \text{ X } 10^3 \text{ J} = 0.387 \text{ J g}^{-1} \, °C^{-1} \times (120 - 26.50) \, °C \times \text{X g}; \quad \text{X} = 34.8 \text{ g}$

5.21   Eating 0.25 lbs of fat is equivalent to 114 g of fat:

$$\# \text{ g fat} = \left(0.25 \text{ lbs fat}\right)\left(\dfrac{1 \text{ kg}}{2.2 \text{ lbs}}\right)\left(\dfrac{1000 \text{ g}}{1 \text{ kg}}\right) = 114 \text{ g fat}$$

This is equivalent to 1.46 lbs of fat tissue:

$$\# \text{ lbs fat tissue} = \left(114 \text{ g fat}\right)\left(\dfrac{9.0 \text{ kcal}}{1 \text{ g fat}}\right)\left(\dfrac{0.50 \text{ lbs fat tissue}}{350 \text{ kcal}}\right) = 1.46 \text{ lbs fat tissue}$$

It will take 1.5 days to lose this amount of fat tissue:

$$\# \text{ days} = \left(1 \text{ kg fat tissue}\right)\left(\dfrac{2.2 \text{ lbs}}{1 \text{ kg}}\right)\left(\dfrac{1 \text{ day}}{1.46 \text{ lbs fat tissue}}\right) = 1.5 \text{ days}$$

5.23 $\dfrac{\# \text{ J}}{\text{mol } °C} = \left(\dfrac{0.586 \text{ cal}}{g \, °C}\right)\left(\dfrac{4.184 \text{ J}}{\text{cal}}\right)\left(\dfrac{46.1 \text{ g C}_2\text{H}_5\text{OH}}{1 \text{ mol C}_2\text{H}_5\text{OH}}\right) = 113 \, ^\ast \text{J} \Big/ \text{mol } °C$

5.25 Keep in mind that the total mass must be considered in this calculation, and that both liquids, once mixed, undergo the same temperature increase:

$$q = (4.18 \text{ J/g °C}) \times (55.0 \text{ g} + 55.0 \text{ g}) \times (31.8 \text{ °C} - 23.5 \text{ °C})$$
$$= 3.8 \times 10^3 \text{ J of heat energy released}$$

Next determine the number of moles of reactant involved in the reaction:
$0.0550 \text{ L} \times 1.3 \text{ mol/L} = 0.072$ mol of acid and of base.

Thus the enthalpy change is: $\dfrac{\# \text{ kJ}}{\text{mol}} = \dfrac{\left(-3.8 \times 10^3 \text{ J}\right)\left(\dfrac{1 \text{ kJ}}{1000 \text{ J}}\right)}{(0.072 \text{ mol})} = -53 \text{ kJ}\big/\text{mol}$

5.27 This problem may be solved exactly as for problem 5.26.

total heat capacity:
$(4.031 \text{ J/g °C} \times 610.28 \text{ g}) + (4.003 \text{ J/g °C} \times 619.69 \text{ g}) + 77.99 \text{ J/°C} = 5019 \text{ J/°C}$

common temperature change: $19.410 \text{ °C} - 15.533 \text{ °C} = 3.877 \text{ °C}$

Heat flow to the system:
$$q = 5019 \text{ J/°C} \times 3.877 \text{ °C} = 1.946 \times 10^4 \text{ J} = 1.946 \times 10^1 \text{ kJ}$$

The enthalpy change is thus:
$$\Delta H = -1.946 \times 10^1 \text{ kJ} \div 0.33143 \text{ mol} = -58.72 \text{ kJ/mol}$$

5.36 Multiply the given equation by the fraction 2/3.
(a) $2CO(g) + O_2(g) \rightarrow 2CO_2(g)$, $\Delta H° = -566 \text{ kJ}$
(b) To determine ΔH for 1 mol, simply multiply the original ΔH by 1/3; -283 kJ/mol.

5.38 $4Al(s) + 2Fe_2O_3(s) \rightarrow 2Al_2O_3(s) + 4Fe(s)$, $\Delta H° = -1708 \text{ kJ}$

5.40 $10CaO(s) + 10H_2O(\ell) \rightarrow 10Ca(OH)_2(s)$, $\Delta H° = -653 \text{ kJ}$

5.42

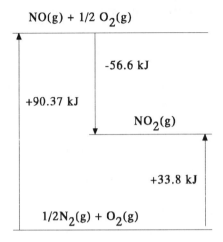

NO(g) + 1/2 O_2(g)

-56.6 kJ

+90.37 kJ

NO_2(g)

+33.8 kJ

1/2N_2(g) + O_2(g)

5.45   Divide the second equation by 2:
       NO(g) + 1/2O_2(g) → NO_2(g), $\Delta H° = -56.570$ kJ

5.47   If we label the four known thermochemical equations consecutively, 1, 2, 3, and 4, then
       the sum is made in the following way: Divide equation #3 by two, and reverse all of the
       other equations (#1, #2, and #4), while also dividing each by two:

1/2Na_2O + HCl	→ 1/2H_2O + NaCl,	$\Delta H° = -253.66$ kJ
$NaNO_2$	→ 1/2Na_2O + 1/2NO_2 + 1/2NO,	$\Delta H° = 213.57$ kJ
1/2NO + 1/2NO_2	→ 1/2N_2O + 1/2O_2,	$\Delta H° = -21.34$ kJ
1/2H_2O + 1/2O_2 + 1/2N_2O	→ HNO_2,	$\Delta H° = -17.18$ kJ

       Adding gives:

HCl + $NaNO_2$	→ NaCl + HNO_2,	$\Delta H° = -78.61$ kJ

5.49   The desired net equation is obtained by adding together the reverse of the two
       thermochemical equations:

       $Zn(NO_3)_2$(aq) + Cu(s) → $Cu(NO_3)_2$(aq) + Zn(s), $\Delta H° = 258$ kJ
       $Cu(NO_3)_2$(aq) + 2Ag(s) → $2AgNO_3$(aq) + Cu(s), $\Delta H° = 106$ kJ

       2Ag(s) + $Zn(NO_3)_2$(aq) → Zn(s) + $2AgNO_3$(aq), $\Delta H° = 364$ kJ

       Since this $\Delta H°$ has a positive value, it is the reverse reaction that occurs spontaneously.

5.51   Reverse the second and the third thermochemical equations and add them to the first:

$$CaO(s) + 2HCl(aq) \rightarrow CaCl_2(aq) + H_2O(\ell), \qquad \Delta H° = -186 \text{ kJ}$$
$$Ca(OH)_2(s) \rightarrow CaO(s) + H_2O(\ell), \qquad \Delta H° = 62.3 \text{ kJ}$$
$$Ca(OH)_2(aq) \rightarrow Ca(OH)_2(s), \qquad \Delta H° = 12.6 \text{ kJ}$$

$$Ca(OH)_2(aq) + 2HCl(aq) \rightarrow CaCl_2(aq) + 2H_2O(\ell), \qquad \Delta H° = -111 \text{ kJ}$$

5.53   Only (b) should be labeled with $\Delta H_f°$.

5.56   (a) $1/2H_2(g) + 1/2Cl_2(g) \rightarrow HCl(g)$, $\Delta H_f° = -92.30 \text{ kJ/mol}$
       (b) $1/2N_2(g) + 2H_2(g) + 1/2Cl_2(g) \rightarrow NH_4Cl(s)$, $\Delta H_f° = -315.4 \text{ kJ/mol}$
       (c) $C(s) + 1/2O_2(g) + 2H_2(g) + N_2(g) \rightarrow CO(NH_2)_2(s)$, $\Delta H_f° = -333.19 \text{ kJ/mol}$
       (d) $2Na(s) + C(s) + 3/2O_2(g) \rightarrow Na_2CO_3(s)$, $\Delta H_f° = -1131 \text{ kJ/mol}$

5.57 $\Delta H° = 3\Delta H_f°[CO_2(g)] + 4\Delta H_f°[Fe(s)] - 2\Delta H_f°[Fe_2O_3(s)] - 3\Delta H_f°[C(s)]$
              $= 3 \text{ mol} \times (-393.5 \text{ kJ/mol}) + 4 \text{ mol} \times (0.0 \text{ kJ/mol})$
                          $- 2 \text{ mol} \times (-822.2 \text{ kJ/mol}) - 3 \text{ mol} \times (0.0 \text{ kJ/mol})$
              $= 463.9 \text{ kJ}$
       This reaction is endothermic.

5.59   These two thermochemical equations are added along with six times that for the formation of liquid water:
       $$P_4O_{10}(s) + 6H_2O(\ell) \rightarrow 4H_3PO_4(\ell), \qquad \Delta H° = -257.2 \text{ kJ}$$
       $$4P(s) + 5O_2(g) \rightarrow P_4O_{10}(s), \qquad \Delta H° = -3062 \text{ kJ}$$
       $$6H_2(g) + 3O_2(g) \rightarrow 6H_2O(\ell), \qquad \Delta H° = -1715.4 \text{ kJ}$$

       Adding gives:

       $$4P(s) + 6H_2(g) + 8O_2(g) \rightarrow 4H_3PO_4(\ell), \qquad \Delta H° = -5035 \text{ kJ}$$

       The above result should then be divided by four:
       $$P(s) + 3/2H_2(g) + 2O_2(g) \rightarrow H_3PO_4(\ell), \qquad \Delta H° = -1259 \text{ kJ/mol}$$

5.61   The equation may be written as: $1/2\ H_2(g) + 1/2\ Br_2(\ell) \rightarrow HBr(g)$; $\Delta H_f^\circ = -36$ kJ

To obtain ΔH, combine the equations in the following manner:

$Br_2(aq) + 2KCl(aq)$	$\rightarrow Cl_2(g) + 2KBr(aq)$	$\Delta H^\circ = 96.2$ kJ
$H_2(g) + Cl_2(g)$	$\rightarrow 2HCl(g)$	$\Delta H^\circ = -184$ kJ
$2HCl(aq) + 2KOH(aq)$	$\rightarrow 2KCl(aq) + 2H_2O(\ell)$	$\Delta H^\circ = -115$ kJ
$2KBr(aq) + 2H_2O(\ell)$	$\rightarrow 2HBr(aq) + 2KOH(aq)$	$\Delta H^\circ = 115$ kJ
$2HCl(g)$	$\rightarrow 2HCl(aq)$	$\Delta H^\circ = -154$ kJ
$2HBr(aq)$	$\rightarrow 2HBr(g)$	$\Delta H^\circ = 160$ kJ
$Br_2(\ell)$	$\rightarrow Br_2(aq)$	$\Delta H^\circ = -4.2$ kJ

Add all of the above to get;

$H_2(g) + Br_2(\ell) \rightarrow 2HBr(g)$; $\Delta H = -86$ kJ

Now divide this equation by two to give the thermochemical equation for the formation of 1 mol of HBr(g):

$1/2\ H_2(g) + 1/2\ Br_2(\ell) \rightarrow HBr(g)$; $\Delta H = -43$ kJ

Comparing this value to the ΔH_f° value listed in Appendix E and at the outset of this problem, we see that this experimental data indicates a value that is close to the reported value.

CHAPTER SIX
Review Exercises

6.9 $\quad \nu = \dfrac{c}{\lambda} = \dfrac{3.00 \times 10^8 \ m/s}{430 \times 10^{-9} \ m} = 6.98 \times 10^{14} \ s^{-1} = 6.98 \times 10^{14} \ Hz$

6.11 $\quad 295 \ nm = 295 \times 10^{-9} \ m$

$\quad \nu = \dfrac{c}{\lambda} = \dfrac{3.00 \times 10^8 \ m/s}{295 \times 10^{-9} \ m} = 1.02 \times 10^{15} \ s^{-1} = 1.02 \times 10^{15} \ Hz$

6.13 $\quad 101.1 \ MHz = 101.1 \times 10^6 \ Hz = 101.1 \times 10^6 \ s^{-1}$

$\quad \lambda = \dfrac{c}{\nu} = \dfrac{3.00 \times 10^8 \ m/s}{101.1 \times 10^6 \ s^{-1}} = 2.98 \ m$

6.15 $\quad \lambda = \dfrac{c}{\nu} = \dfrac{3.00 \times 10^8 \ m/s}{60 \ s^{-1}} = 5.0 \times 10^6 \ m = 5.0 \times 10^3 \ km$

6.22 $\quad E = h\nu = 6.63 \times 10^{-34} \ J \ s \times 4.0 \times 10^{14} \ s^{-1} = 2.7 \times 10^{-19} \ J$

$\quad \dfrac{\# \ J}{mol} = \left(\dfrac{2.7 \times 10^{-19} \ J}{1 \ photon} \right)\left(\dfrac{6.022 \times 10^{23} \ photons}{1 \ mol} \right) = 1.6 \times 10^5 \ J \ mol^{-1}$

6.24 We proceed by calculating the wavelength of a single photon:

$$E = \dfrac{hc}{\lambda} = \dfrac{\left(6.626 \times 10^{-34} \ J \ s \right)\left(3.00 \times 10^8 \ m/s \right)}{\left(3.00 \times 10^{-3} \ m \right)} = 6.63 \times 10^{-23} \ J$$

Since the specific heat of water is 4.184 J/g °C, it will take 4.184 J of heat energy to raise the temperature of the water. The required number of photons is then:

$$4.184 \ J \times \dfrac{1 \ photon}{6.63 \times 10^{-23} \ J} = 6.32 \times 10^{22} \ photons$$

6.27 (a) violet

(b) $\nu = c/\lambda = 3.00 \times 10^{8} \text{ m s}^{-1} \div 410.3 \times 10^{-9} \text{ m} = 7.31 \times 10^{14} \text{ s}^{-1}$

(c) $E = h\nu = 6.63 \times 10^{-34} \text{ J s} \times 7.31 \times 10^{14} \text{ s}^{-1} = 4.85 \times 10^{-19} \text{ J}$

6.28

$$\frac{1}{\lambda} = 109{,}678 \text{ cm}^{-1} \times \left(\frac{1}{3^2} - \frac{1}{6^2}\right) = 109{,}678 \text{ cm}^{-1} \times (0.1111 - 0.02778)$$

$$\frac{1}{\lambda} = 9.140 \times 10^{3} \text{ cm}^{-1}$$

$\lambda = 1.094 \times 10^{-4} \text{ cm} = 1090 \text{ nm}$, which is not in the the visible region.

6.32

$$\frac{1}{\lambda} = 109{,}678 \text{ cm}^{-1} \times \left(\frac{1}{1^2} - \frac{1}{4^2}\right) = 109{,}678 \text{ cm}^{-1} \times (1 - 0.0625)$$

$$\frac{1}{\lambda} = 102{,}823 \text{ cm}^{-1}$$

$\lambda = 9.73 \times 10^{-6} \text{ cm} = 97.3 \text{ nm}$, which is in the ultraviolet region.

$E = hc/\lambda$

$$E = \frac{hc}{\lambda} = \frac{\left(6.626 \times 10^{-34} \text{ J s}\right)\left(3.00 \times 10^{8} \text{ m}/\text{s}\right)}{\left(97.3 \times 10^{-9} \text{ m}\right)} = 2.04 \times 10^{-18} \text{ J}$$

6.32

$$\frac{1}{\lambda} = 109{,}678 \text{ cm}^{-1} \times \left(\frac{1}{1^2} - \frac{1}{4^2}\right) = 109{,}678 \text{ cm}^{-1} \times (1 - 0.0625)$$

$$\frac{1}{\lambda} = 102{,}823 \text{ cm}^{-1}$$

$\lambda = 9.73 \times 10^{-6} \text{ cm} = 97.3 \text{ nm}$, which is in the ultraviolet region.

$E = hc/\lambda$

$$E = \frac{hc}{\lambda} = \frac{\left(6.626 \times 10^{-34} \text{ J s}\right)\left(3.00 \times 10^8 \text{ }^m/_s\right)}{\left(97.3 \times 10^{-9} \text{ m}\right)} = 2.04 \times 10^{-18} \text{ J}$$

6.43 (a) $n = 1$ (b) $n = 3$

6.45 (a) $n = 3, \ell = 0$ (b) $n = 5, \ell = 2$ (c) $n = 4, \ell = 3$

6.48 (a) $m_\ell = 1, 0, \text{ or } -1$ (b) $m_\ell = 3, 2, 1, 0, -1, -2, \text{ or } -3$

6.50 No, the maximum value of ℓ in the 4th shell ($n = 4$) is 3.

6.52 When $m_\ell = -4$ the minimum value of ℓ is 4 and the minimum value of n is 5.

6.54 There are eleven values for m_ℓ: $-5, -4, -3, -2, -1, 0, 1, 2, 3, 4,$ and 5. Thus there are eleven orbitals.

6.65 (a) S $1s^2 2s^2 2p^6 3s^2 3p^4$
      (b) K $1s^2 2s^2 2p^6 3s^2 3p^6 4s^1$
      (c) Ti $1s^2 2s^2 2p^6 3s^2 3p^6 4s^2 3d^2$
      (d) Sn $1s^2 2s^2 2p^6 3s^2 3p^6 4s^2 3d^{10} 4p^6 5s^2 4d^{10} 5p^2$

6.69

   (a) Mg: ⚛ ⚛ ⚛⚛⚛ ⚛ ◯◯◯ ◯ ◯◯◯◯◯
               1s 2s 2p 3s 3p 4s 3d

   (b) Ti: ⚛ ⚛ ⚛⚛⚛ ⚛ ⚛⚛⚛ ⚛ ⬆⬆◯◯◯
               1s 2s 2p 3s 3p 4s 3d

6.71 (a) Mg is $1s^2 2s^2 2p^6 3s^2$, $\therefore$ zero unpaired electrons
      (b) P is $1s^2 2s^2 2p^6 3s^2 3p^3$, $\therefore$ three unpaired electrons
      (c) V is $1s^2 2s^2 2p^6 3s^2 3p^6 4s^2 3d^3$, $\therefore$ three unpaired electrons

6.72 (a) Ni $[Ar]4s^2 3d^8$ or $[Ar]3d^8 4s^2$

(b) Cs $[Xe]6s^1$

(c) Ge $[Ar]4s^2 3d^{10} 4p^2$ or $[Ar]3d^{10} 4s^2 4p^2$

(d) Br $[Ar]4s^2 3d^{10} 4p^5$ or $[Ar]3d^{10} 4s^2 4p^5$

6.73

(a) Ni: [Ar] ⊛ ⊛⊛⊛⊛⊛
4s   3d

(b) Cs: [Xe] ⊛
6s

(c) Ge: [Ar] ⊛ ⊛⊛⊛⊛⊛ ⊛⊛○
4s   3d   4p

(d) Br: [Ar] ⊛ ⊛⊛⊛⊛⊛ ⊛⊛⊛
4s   3d   4p

6.78 The value corresponds to the row in which the element resides:
(a) 5 (b) 4 (c) 4 (d) 6

6.79 (a) Na $3s^1$ (b) Al $3s^2 3p^1$ (c) Ge $4s^2 4p^2$ (d) P $3s^2 3p^3$

6.80

(a) Na: ⊛
3s

(c) Ge: ⊛ ⊛⊛○
4s 4p

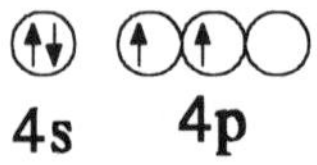

(b) Al: ⊛ ⊛○○
3s 3p

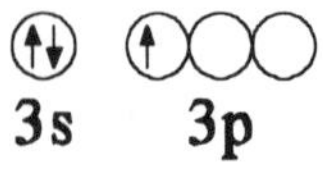

(d) P: ⊛ ⊛⊛⊛
3s 3p

54

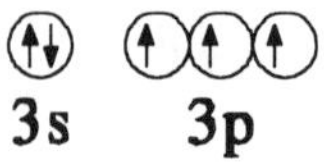

6.92   (a)   1   (b)   6   (c)   7

6.93   (a)   Na   (b)   Sb

6.95   Sb is a bit larger than Sn, according to Figure 6.24.

6.97   Since these atoms and ions all have the same number of electrons, the size should be inversely related to the positive charge:

$$Mg^{2+} < Na^{+} < Ne < F^{-} < O^{2-} < N^{3-}$$

6.99   Cations are generally smaller than the corresponding atom, and anions are generally larger than the corresponding atom:

   (a)   Na   (b) Co^{2+} (c) Cl^{-}

6.102   (a)   C   (b)   O   (c)   Cl

6.111   (a)   Cl   (b)   Br   (c)   Si

6.114   (a)   We must first calculate the energy in joules of a mole of photons.

$$E = \frac{hc}{\lambda} = \frac{\left(6.63 \times 10^{-34} \text{ J s}\right)\left(3.00 \times 10^{8} \text{ m/s}\right)}{600 \times 10^{-9} \text{ m}} = 3.32 \times 10^{-19} \text{ J / photon}$$

3.32×10^{-19} J / photon $\times$ 6.02×10^{23} photons / mol $=$ 2.00×10^{5} J / mol

Next, we calculate the heat transfer problem as in Chapter 4:

2.00×10^{5} J $= 4.18$ J g^{-1} °C^{-1} $\times$ X g $\times$ 5.0 °C
$X = 9.57 \times 10^{3}$ g

   (b) $E = 6.63 \times 10^{-19}$ J/photon
         $E = 3.99 \times 10^{5}$ J/mol
         $X = 1.91 \times 10^{4}$ g

6.116 (a)

$$\frac{1}{\lambda} = 109{,}678 \text{ cm}^{-1} \times \left(\frac{1}{2^2} - \frac{1}{5^2}\right) = 109{,}678 \text{ cm}^{-1} \times (0.2500 - 0.04000)$$

$$\frac{1}{\lambda} = 2.303 \times 10^4 \text{ cm}^{-1}$$

$\lambda = 4.342 \times 10^{-5}$ cm = 434.2 nm, which is in the violet region of the visible spectrum.

(b)

$$\frac{1}{\lambda} = 109{,}678 \text{ cm}^{-1} \times \left(\frac{1}{1^2} - \frac{1}{4^2}\right) = 109{,}678 \text{ cm}^{-1} \times (1 - 0.0625)$$

$$\frac{1}{\lambda} = 1.028 \times 10^5 \text{ cm}^{-1}$$

$\lambda = 9.725 \times 10^{-6}$ cm = 97.25 nm, which is in the ultraviolet region of the spectrum.

(c)

$$\frac{1}{\lambda} = 109{,}678 \text{ cm}^{-1} \times \left(\frac{1}{4^2} - \frac{1}{6^2}\right) = 109{,}678 \text{ cm}^{-1} \times (0.0625 - 0.0278)$$

$$\frac{1}{\lambda} = 3.808 \times 10^3 \text{ cm}^{-1}$$

$\lambda = 2.626 \times 10^{-4}$ cm = 2628 nm, which is in the infrared region of the spectrum.

6.118 (a)   This diagram violates the aufbrau principle. Specifically, the s-orbital should be filled before filling the higher energy p-orbitals.

     (b)   This diagram violates the aufbrau principle. Specifically, the lower energy s-orbital should be filled completely before filling the p-orbitals.

     (c)   This diagram violates the aufbrau principle. Specifically, the lower energy s-orbital should be filled completely before filling the p-orbitals.

6.120   The 4s electrons are lost; $n = 4$, $\ell = 0$, $m_\ell = 0$, $m_s = \pm 1/2$

6.124 We simply reverse the electron affinities of the corresponding ions.

(a) $F^-(g) \rightarrow F(g) + e^-$, $\Delta H^\circ = 328$ kJ/mol

(b) $O^-(g) \rightarrow O(g) + e^-$, $\Delta H^\circ = 141$ kJ/mol

(c) $O^{2-}(g) \rightarrow O^-(g) + e^-$, $\Delta H^\circ = -844$ kJ/mol

The last of these is exothermic, meaning that loss of an electron from the oxide ion is favorable from the standpoint of enthalpy.

7.5   Magnesium loses two electrons:

$$Mg \quad \rightarrow \quad Mg^{2+} + 2e^-$$
$$[Ne]3s^2 \qquad [Ne]$$

Bromine gains an electron:

$$Br \quad + e^- \rightarrow \quad Br^-$$
$$[Ar]3d^{10}4s^24p^5 \qquad [Kr]$$

To keep the overall charge of the molecule neutral, two Br^- ions combine with one Mg^{2+} ion to form $MgBr_2$:

$$Mg^{2+} + 2Br^- \rightarrow MgBr_2$$

7.9 Pb^{2+}: $[Xe]6s^24f^{14}5d^{10}$
       Pb^{4+}: $[Xe]4f^{14}5d^{10}$

7.11   (a) $[Ar]3d^8$ (b) $[Ar]3d^2$ (c) $[Ar]$
       (d) $[Ar]3d^9$ (e) $[Xe]5d^8$

7.13

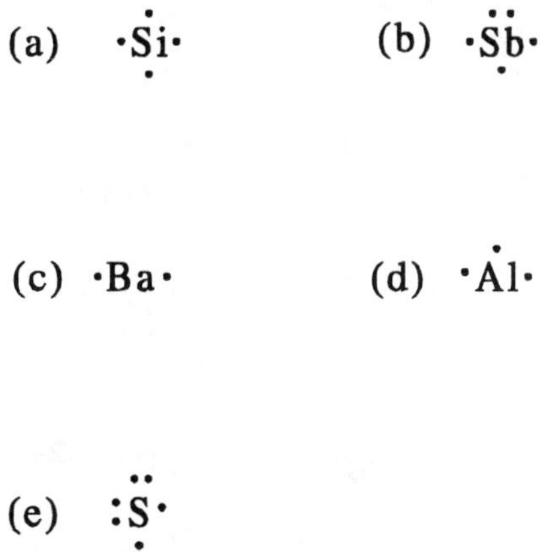

(a) ·S̈i·   (b) ·S̈b·

(c) ·Ba·   (d) ·A̋l·

(e) :S̈·

7.15

(a) [K] $^+$ (b) [Al] $^{3+}$

(c) [:S̈:] $^{2-}$ (d) [:S̈i:] $^{4-}$

(e) [Mg] $^{2+}$

7.17

(a)

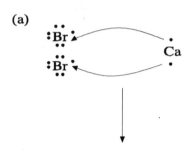

2[:B̈r:] $^-$ +   [Ca] $^{2+}$

(b)

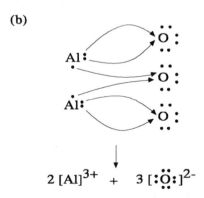

2 [Al] $^{3+}$ + 3 [:Ö:] $^{2-}$

(c)

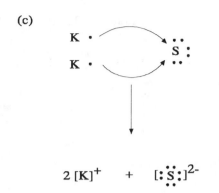

$$2\,[\mathrm{K}]^{+} \quad + \quad [:\!\overset{\cdot\cdot}{\underset{\cdot\cdot}{\mathrm{S}}}\!:]^{2-}$$

7.26 $\#\mathrm{J\,/\,molecule} = (435 \times 10^{3}\ \mathrm{J\,/\,mole})\left(\dfrac{1\ \mathrm{mole}}{6.02 \times 10^{23}\ \mathrm{molecules}}\right)$

$= 7.23 \times 10^{-19}\ \mathrm{J\,/\,molecule}$

7.28

(a) $:\!\overset{\cdot\cdot}{\underset{\cdot\cdot}{\mathrm{Br}}}\!\cdot \;+\; \cdot\overset{\cdot\cdot}{\underset{\cdot\cdot}{\mathrm{Br}}}\!: \;\longrightarrow\; :\!\overset{\cdot\cdot}{\underset{\cdot\cdot}{\mathrm{Br}}}\!-\!\overset{\cdot\cdot}{\underset{\cdot\cdot}{\mathrm{Br}}}\!:$

(b) $2\,\mathrm{H}\cdot \;+\; \cdot\overset{\cdot\cdot}{\underset{\cdot\cdot}{\mathrm{O}}}\!\cdot \;\longrightarrow\; \mathrm{H}\!-\!\overset{\cdot\cdot}{\underset{\cdot\cdot}{\mathrm{O}}}\!:$
        $|$
        H

(c) $3\,\mathrm{H}\cdot \;+\; \cdot\overset{\cdot\cdot}{\underset{\cdot}{\mathrm{N}}}\!\cdot \;\longrightarrow\; \mathrm{H}\!-\!\overset{\cdot\cdot}{\mathrm{N}}\!-\!\mathrm{H}$
        $|$
        H

7.30 (a) one (b) four (c) two
      (d) three (e) one

7.32 (a) We predict the formula H_2Se because selenium, being in Group VIA, needs only two additional electrons (one each from two hydrogen atoms) in order to complete its octet.

      (b) Arsenic, being in Group VA, needs three electrons from hydrogen atoms in order to complete its octet, and we predict the formula H_3As.

      (c) Silicon is in Group IVB, and it needs four electrons (and hence four hydrogen atoms) to complete its octet: SiH_4.

7.41   This compound has twelve hydrogen atoms and has the formula C_5H_{12}.

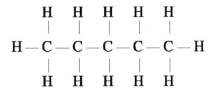

7.44   Like most organic acids, this acid is weak.

$$CH_3CH_2\overset{\overset{\displaystyle O}{\|}}{C}OH \rightleftharpoons CH_3CH_2\overset{\overset{\displaystyle O}{\|}}{C}O^- + H^+$$

7.46

$$CH_3H_2\overset{\overset{\displaystyle O}{\|}}{C}OH + CH_3\overset{\overset{\displaystyle H}{|}}{N}CH_2CH_3 \longrightarrow$$

$$CH_3CH_2\underset{\underset{\displaystyle CH_3}{|}}{\overset{\overset{\displaystyle O}{\|}}{C}}NCH_2CH_3 + H_2O$$

7.51   The noble gases are assigned electronegativity values of zero because they have a complete shell of electrons.

7.52   Here we choose the atom with the smaller electronegativity:
(a)   S   (b)   Si   (c)   Br

7.53   Here we choose the linkage that has the greatest difference in electronegativities between the atoms of the bond: N—S.

7.57   (a)   4 electrons or two bonding pairs
(b)   six electrons, all in bonding pairs
(c)   two electrons for each hydrogen

7.59

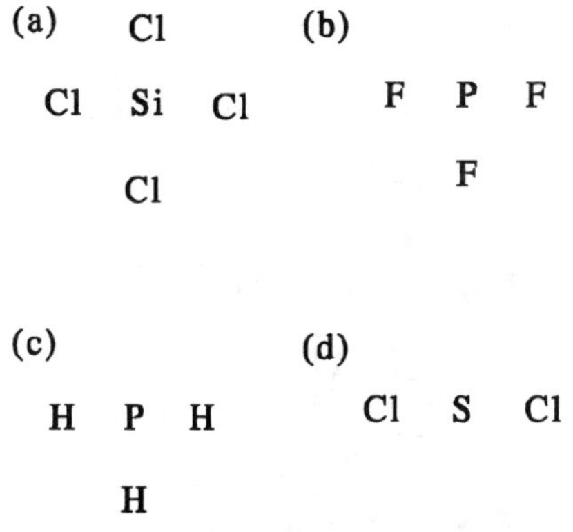

(a)

$$Cl$$
$$Cl \quad Si \quad Cl$$
$$Cl$$

(b)

$$F \quad P \quad F$$
$$F$$

(c)

$$H \quad P \quad H$$
$$H$$

(d)

$$Cl \quad S \quad Cl$$

7.61 (a)   32   (b)   26   (c)   8   (d)   20

7.63

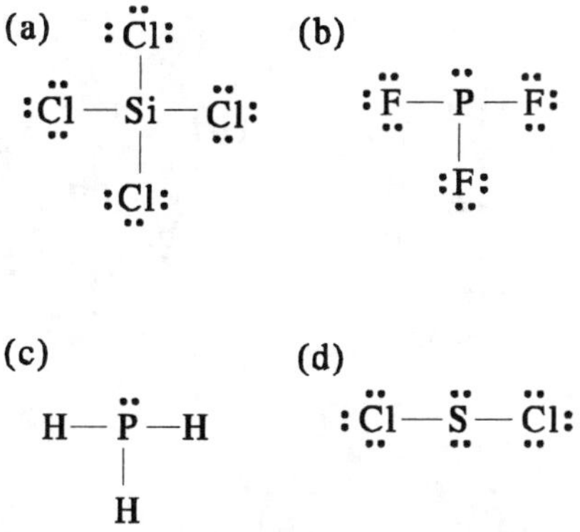

(a)

$$
:\!\overset{\displaystyle ..}{\underset{\displaystyle ..}{Cl}}\!:
$$
$$
:\!\overset{..}{\underset{..}{Cl}}\!-Si-\overset{..}{\underset{..}{Cl}}\!:
$$
$$
:\!\overset{..}{\underset{..}{Cl}}\!:
$$

(b)

$$
:\!\overset{..}{\underset{..}{F}}\!-P-\overset{..}{\underset{..}{F}}\!:
$$
$$
:\!\overset{..}{\underset{..}{F}}\!:
$$

(c)

$$
H-\overset{..}{P}-H
$$
$$
H
$$

(d)

$$
:\!\overset{..}{\underset{..}{Cl}}\!-\overset{..}{\underset{..}{S}}\!-\overset{..}{\underset{..}{Cl}}\!:
$$

7.65

(a)

$$
\overset{.}{\underset{..}{S}}\!=C=\overset{.}{\underset{..}{S}}\!:
$$

(b)

$$
\left[:\!C\!\equiv\!N\!:\right]^{-}
$$

(c)

$$
:\!\overset{..}{\underset{..}{O}}\!-Se-\overset{..}{\underset{..}{O}}\!:
$$
$$
\parallel
$$
$$
.\overset{.}{O}.
$$

(d)

$$
:\!\overset{..}{\underset{..}{O}}\!-Se=\overset{..}{O}\!:
$$

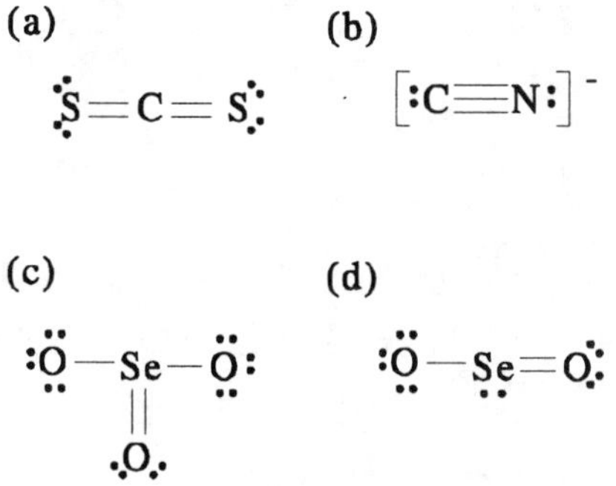

7.66

(a)

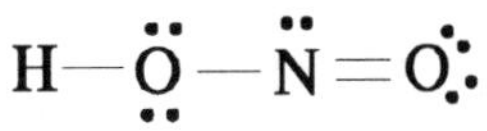

$$\text{H}-\overset{..}{\underset{..}{\text{O}}}-\overset{..}{\text{N}}=\overset{..}{\text{O}}:$$

(b)

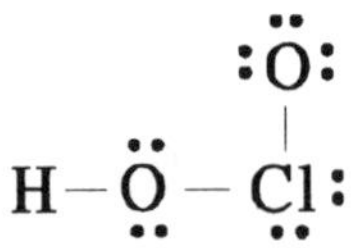

$$\text{H}-\overset{..}{\underset{..}{\text{O}}}-\underset{..}{\text{Cl}}-\overset{\overset{..}{\text{O}}:}{\underset{|}{}}$$

(c)

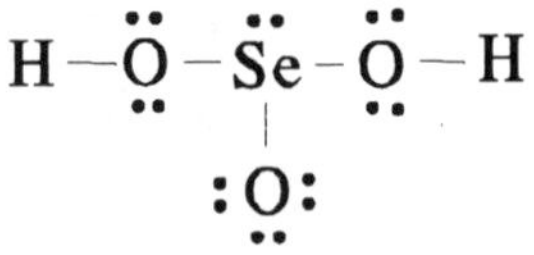

$$\text{H}-\overset{..}{\underset{..}{\text{O}}}-\overset{..}{\underset{\overset{|}{:\text{O}:}}{\text{Se}}}-\overset{..}{\underset{..}{\text{O}}}-\text{H}$$

7.68

(a) TeF_4 with $:\text{Te}:$ bonded to four F

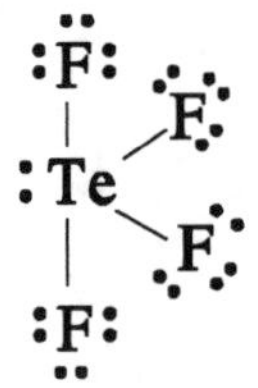

(b) ClF_5 with Cl bonded to five F

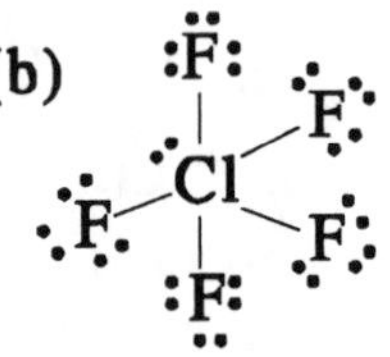

(c) XeF_2 with $:\text{Xe}:$ bonded to two F

(d) XeF_4 with Xe bonded to four F

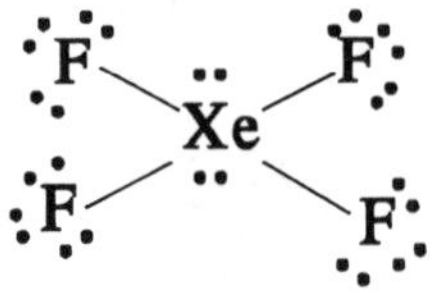

7.77

(a)

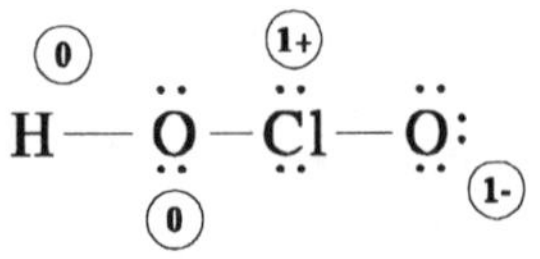

$$\text{H}-\overset{(0)}{\overset{..}{\text{O}}}-\overset{(1+)}{\underset{(0)}{\text{Cl}}}-\overset{(1-)}{\overset{..}{\text{O}}}:$$

(b)

$$\overset{(1-)}{\underset{(0)\;\;\overset{..}{\text{O}}=\overset{(2+)}{\text{S}}-\overset{..}{\text{O}}:\;(1-)}{:\text{O}:}}$$

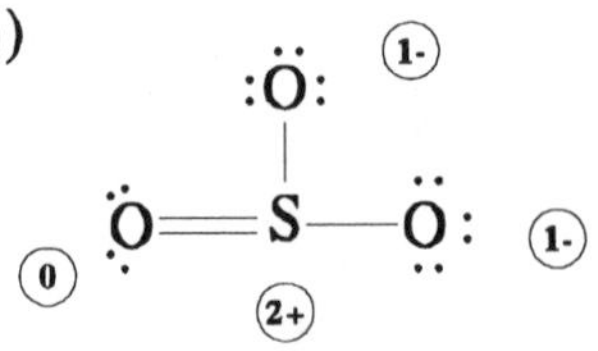

(c)

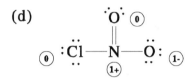

(d)

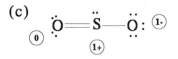

(e)

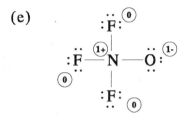

7.79   The formal charges on all of the atoms of the left structure are zero, therefore, the potential energy of this molecule is lower and it is more stable.

7.81

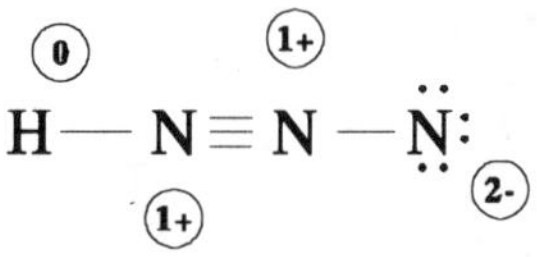

$$H — N \equiv N — \ddot{N}:$$

7.85   The Lewis structure for NO_3^- is given in the answer to practice exercise 7, and that for NO_2^- is given in the answer to review exercise 7.64.

Resonance causes the average number of bonds in each N—O linkage of NO_3^- to be 1.33. Resonance causes the average number of electron pair bonds in each linkage of NO_2^- to be 1.5. We conclude that the N—O bond in NO_2^- should be shorter than that in NO_3^-.

7.88

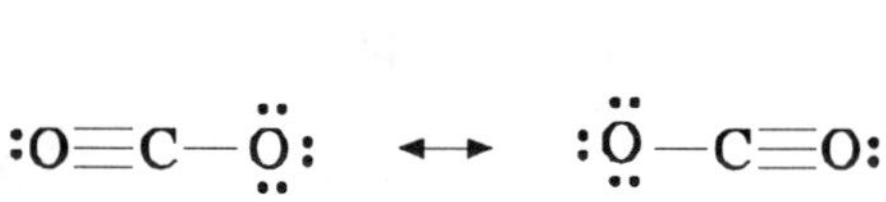

$$:O \equiv C — \ddot{O}: \quad \longleftrightarrow \quad :\ddot{O} — C \equiv O:$$

These are not preferred structures, because in each Lewis diagram, one oxygen bears a formal charge of +1 whereas the other bears a formal charge of –1. The structure with the formal charges of zero has a lower potential energy and is more stable.

7.90   The Lewis structure that is obtained using the rules of Figure 7.7 is:

Formal charges are indicated. The average bond order is 1.5.

7.93

7.95

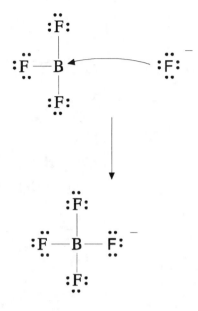

7.98

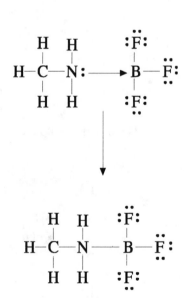

7.100

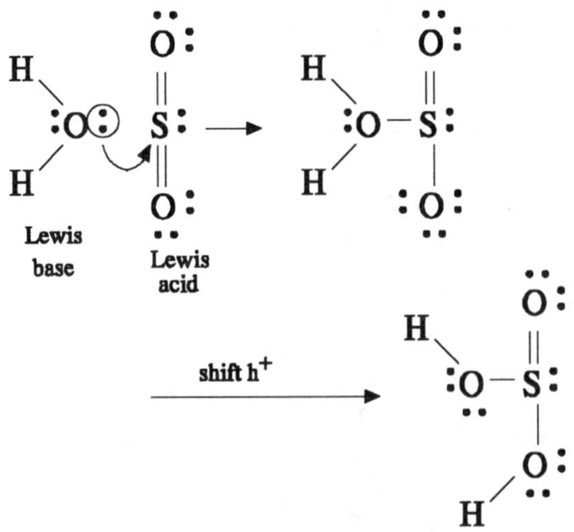

7.102

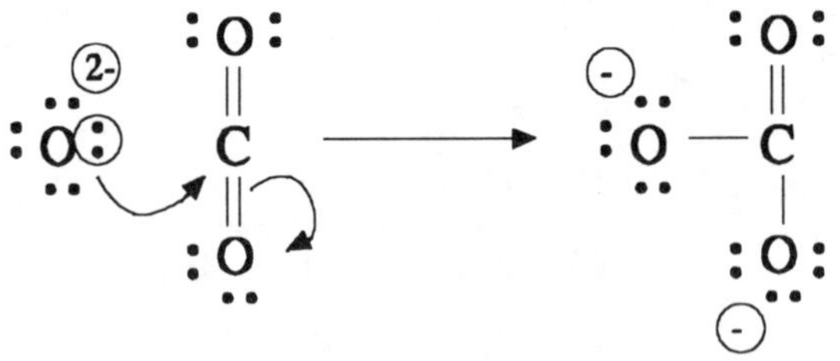

7.109 Only the top two are acceptable. The others violate the octet rule.

7.111

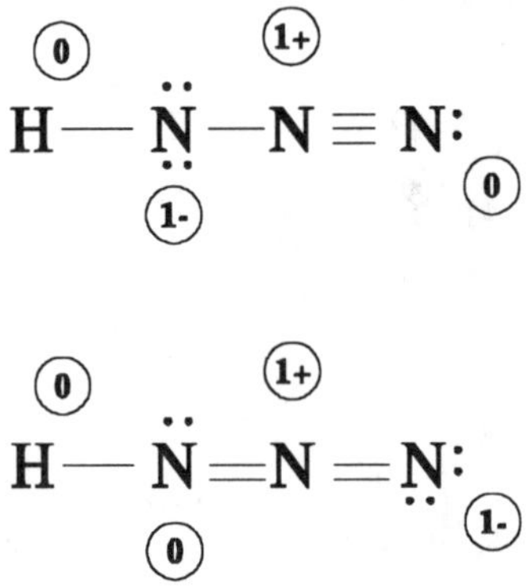

7.113

8.6   (a)   nonlinear   (b)   trigonal bipyramidal
        (c)   pyramidal   (d)   nonlinear

8.8   (a)   tetrahedral   (b)   square planar
        (c)   octahedral   (d)   linear

8.10   (a)   pyramidal   (b)   tetrahedral
        (c)   pyramidal   (d)   tetrahedral

8.12   all angles 120°

8.18   The ones that are polar are (a), (b), and (c). The last two have symmetrical structures, and although individual bonds in these substances are polar bonds, the geometry of the bonds serves to cause the individual dipole moments of the various bonds to cancel one another.

8.27   The 1s atomic orbitals of the hydrogen atoms overlap with the mutually perpendicular p atomic orbitals of the selenium atom.

Se atom in H_2Se (x = H electron):

              4s   4p

8.32   Elements in period two do not have a d subshell in the valence level.

8.35   (a)   There are three bonds to the central Cl atom, plus one lone pair of electrons. The geometry of the molecule is tetrahedral so the Cl atom is to be sp^3 hybridized:

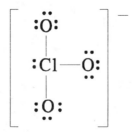

(b)   There are three atoms bonded to the central sulfur atom, and no lone pairs on the central sulfur. The geometry of the molecule is that of a planar triangle, and the hybridization of the S atom is sp^2:

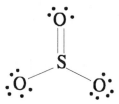

Two other resonance structures should also be drawn for SO_3.

(c)   There are two bonds to the central O atom, as well as two lone pairs. The O atom is to be sp^3 hybridized, and the geometry of the molecule is nonlinear (bent).

(d)   Six Cl atoms surround the central Sb atom in an octahedral geometry, and the hybridization of Sb is sp^3d^2.

(e)   Three Cl atoms are bonded to the Br atom, plus the Br atom has two lone pairs of electrons. This requires the Br atom to be sp^3d hybridized, and the geometry is T–shaped.

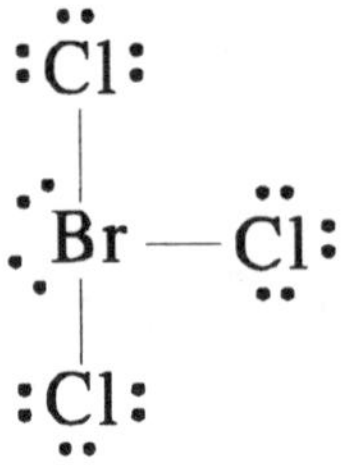

$$:\overset{\displaystyle ..}{Cl}:$$
$$|$$
$$:\overset{\displaystyle .}{\underset{\displaystyle .}{Br}} - \overset{\displaystyle ..}{\underset{\displaystyle ..}{Cl}}:$$
$$|$$
$$:\underset{\displaystyle ..}{Cl}:$$

(f)   The central Xe atom is bonded to four F atoms, plus it has two lone pairs of electrons. The molecule has an octahedral geometry. This requires sp^3d^2 hybridization of Xe.

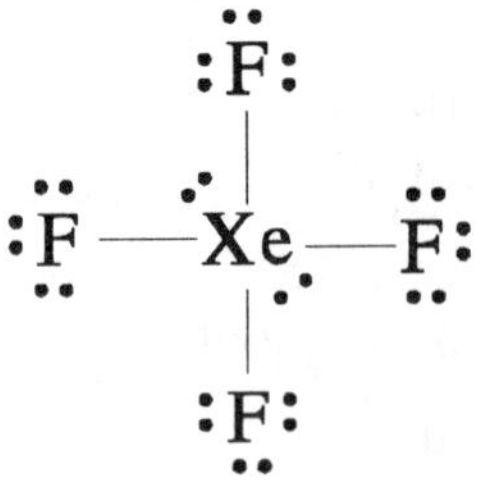

$$:\overset{\displaystyle ..}{F}:$$
$$|$$
$$:\overset{\displaystyle ..}{\underset{\displaystyle ..}{F}} - \overset{\displaystyle .}{\underset{\displaystyle .}{Xe}} - \overset{\displaystyle ..}{\underset{\displaystyle ..}{F}}:$$
$$|$$
$$:\underset{\displaystyle ..}{F}:$$

8.37   (a)   There are three bonds to As and one lone pair at As, requiring As to be sp^3 hybridized.

The Lewis diagram:

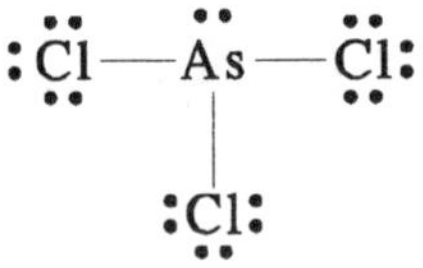

$$:\overset{\displaystyle ..}{\underset{\displaystyle ..}{Cl}} - \overset{\displaystyle ..}{As} - \overset{\displaystyle ..}{\underset{\displaystyle ..}{Cl}}:$$
$$|$$
$$:\underset{\displaystyle ..}{Cl}:$$

The hybrid orbital diagram for As:

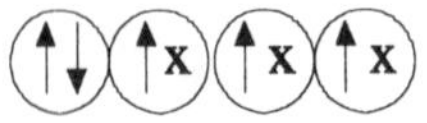

⊛(↑↓)(↑x)(↑x)(↑x)

sp^3

(x = a Cl electron)

(b)   There are three atoms bonded to the central Cl atom, and it also has two lone pairs of electrons. The hybridization of Cl is thus sp^{3}d.

The Lewis diagram:

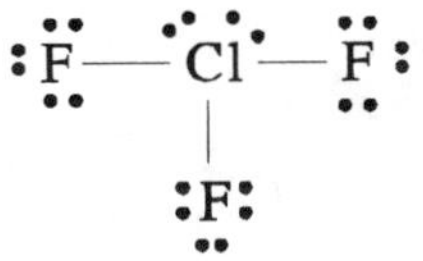

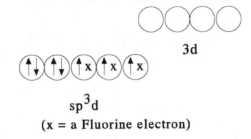

The hybrid orbital diagram for Cl:

3d

sp^{3}d
(x = a Fluorine electron)

(c)   Five Cl atoms are bonded to the central Sb atom, which is therefore sp^{3}d hybridized.

The Lewis diagram:

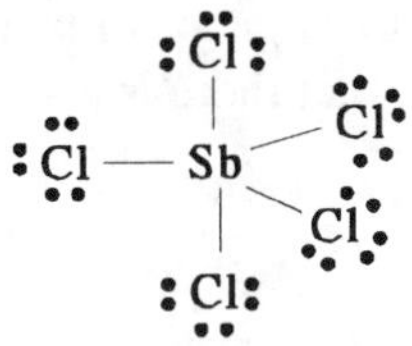

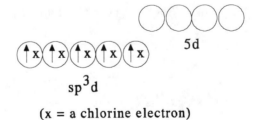

The hybrid orbital diagram for Sb:

5d

sp^{3}d
(x = a chlorine electron)

(d) Se has two bonds to Cl atoms and two lone pairs of electrons, requiring it to be sp^3 hybridized.

The Lewis diagram:

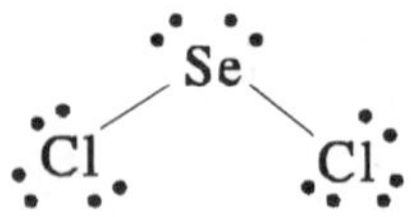

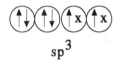

The hybrid orbital diagram for Se:

$$sp^3$$

(x = a chlorine electron)

8.41 The normal C–C–C angle for an sp^3 hybridized carbon atom is 109.5°. The 60° bond angle in cyclopropane is much less than this optimum bond angle. This means that the bonding within the ring cannot be accomplished through the desirable "head on" overlap of hybrid orbitals from each C atom. As a result, the overlap of the hybrid orbitals in cyclopropane is less effective than that in the more normal, noncyclic propane molecule, and this makes the C–C bonds in cyclopropane comparatively weaker than those in the noncyclic molecule. We can also say that there is a severe "ring strain" in the molecule.

8.45 90°

8.48 This is an octahedral ion with sp^3d^2 hybridized tin:

Sn atom in $SnCl_6^{2-}$ (x = Cl electron):

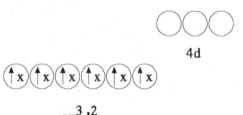

4d

$$sp^3d^2$$

8.54   Whereas the arrangement around an sp^2 hybridized atom may have 120° bond angles (which well accommodates a six–membered ring), the geometrical arrangement around an atom that is sp hybridized must be linear, as in C – C ≡ C – C systems.

8.56   (a)

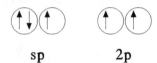

sp-hybridized N atom:

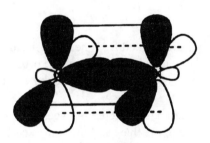

(b)

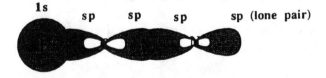

(c)   The σ bonds:

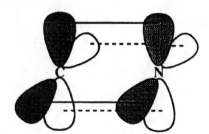

The π bonds:

(d)   The HCN bond angle should be 180°.

8.58   Each carbon atom is sp^2 hybridized, and each C–Cl bond is formed by the overlap of an sp^2 hybrid of carbon with a p atomic orbital of a chlorine atom. The C=C double bond consists first of a C–C σ bond formed by "head on" overlap of sp^2 hybrids from each C atom. Secondly, the C=C double bond consists of a side–to–side overlap of unhybridized p orbitals of each C atom, to give one π bond. The molecule is planar, and the expected bond angles are all 120°.

8.64   As shown in Table 8.1, the bond order of Li_2 is 1.0. The bond order of Be_2 would be zero. Yes, Be_2^+ could exist since the bond order would be 1/2.

8.65   Here we pick the one with the higher bond order.
        (a) O_2^+ (b) O_2 (c) N_2

8.71   (a)   The C–C single bonds are formed from head–to–head overlap of C atom sp^2 hybrids. This leaves one unhybridized atomic p orbital on each carbon atom, and each such atomic orbital is oriented perpendicular to the plane of the molecule.

        (b)   Sideways or π type overlap is expected between the first and the second carbon atoms, as well as between the third and the fourth carbon atoms. However, since all of these atomic p orbitals are properly aligned, there can be continuous π type overlap between all four carbon atoms.

        (c)   The situation described in part (b) is delocalized. We expect completely delocalized π type bonding among the carbon atoms.

        (d)   The bond is shorter because of the extra stability associated with the delocalization energy.

8.81   The 4s level in calcium is filled, but the vacant 4p conduction band overlaps and provides for conductivity.

8.83   Since germanium is in Group IV, an element from Group III must be added to make a p–type semiconductor. Here we could use boron, aluminum, or gallium.

8.88 The double bonds are predicted to be between S and O atoms. Hence, the Cl–S–Cl angle diminishes under the influence of the S=O double bonds.

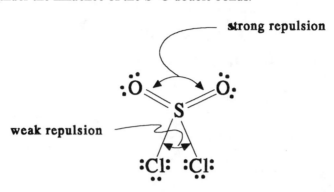

8.90 1. sp^3 2. sp   3. sp^2 4. sp^2

8.91 1. one σ bond
      2. one σ bond and two π bonds
      3. one σ bond
      4. one σ bond and one π bond

8.93 Only the p_y orbital can form a π bond with d_{xy}, if the internuclear axis is the x axis.

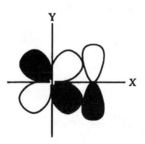

9.2　(a)　HF　(b)　$N_2H_5^+$　(c)　$C_5H_5NH^+$

(d)　HO_2^-　(e)　H_2CrO_4

9.4　(a)

conjugate pair

$$HNO_3 + N_2H_4 \rightleftharpoons N_2H_5^+ + NO_3^-$$
acid　　base　　acid　　base

conjugate pair

(b)

conjugate pair

$$N_2H_5^+ + NH_3 \rightleftharpoons NH_4^+ + N_2H_4$$
acid　　base　　acid　　base

conjugate pair

(c)

conjugate pair

$$H_2PO_4^- + CO_3^{2-} \rightleftharpoons HCO_3^- + HPO_4^{2-}$$
acid　　base　　acid　　base

conjugate pair

(d)

conjugate pair

$$HIO_3 + HC_2O_4^- \rightleftharpoons H_2C_2O_4 + IO_3^-$$
acid　　base　　acid　　base

conjugate pair

9.7

conjugate pair

$$HOCl(aq) + H_2O \rightleftharpoons H_3O^+(aq) + OCl^-(aq)$$
acid　　base　　acid　　base

conjugate pair

OCl^- is a stronger base than water and H_3O^+ is a stronger acid than HOCl

78

9.9 $C_2H_3O_2^-$ is a stronger base than NO_2^-.

9.11   (a) H_2Se (b)   HI   (c) HIO_4 (d) $HClO_4$

9.14   (a)   The relative strength of the binary acids increases from top to bottom in a group of the periodic table, so we expect HAt to be a stronger acid than HI.
       (b) $HAtO_4$

9.19   (a)   ionic: $Mn(s) + 2H^+(aq) + SO_4^{2-}(aq) \rightarrow Mn^{2+}(aq) + H_2(g) + SO_4^{2-}(aq)$
             net ionic: $Mn(s) + 2H^+(aq) \rightarrow Mn^{2+}(aq) + H_2(g)$
       (b)   ionic: $Cd(s) + 2H^+(aq) + SO_4^{2-}(aq) \rightarrow Cd^{2+}(aq) + H_2(g) + SO_4^{2-}(aq)$
             net ionic: $Cd(s) + 2H^+(aq) \rightarrow Cd^{2+}(aq) + H_2(g)$
       (c)   ionic: $Sn(s) + 2H^+(aq) + SO_4^{2-}(aq) \rightarrow Sn^{2+}(aq) + H_2(g) + SO_4^{2-}(aq)$
             net ionic: $Sn(s) + 2H^+(aq) \rightarrow Sn^{2+}(aq) + H_2(g)$
       (d)   ionic: $Ni(s) + 2H^+(aq) + SO_4^{2-}(aq) \rightarrow Ni^{2+}(aq) + H_2(g) + SO_4^{2-}(aq)$
             net ionic: $Ni(s) + 2H^+(aq) \rightarrow Ni^{2+}(aq) + H_2(g)$
       (e)   ionic: $2Cr(s) + 6H^+(aq) + 3SO_4^{2-}(aq) \rightarrow 2Cr^{3+}(aq) + 3H_2(g) + 3SO_4^{2-}(aq)$
             net ionic: $2Cr(s) + 6H^+(aq) \rightarrow 2Cr^{3+}(aq) + 3H_2(g)$

9.21   (a) $3Ag(s) + 4HNO_3(aq) \rightarrow 3AgNO_3(aq) + 2H_2O(\ell) + NO(g)$
       (b) $Ag(s) + 2HNO_3(aq) \rightarrow AgNO_3(aq) + H_2O(\ell) + NO_2(aq)$

9.23 $Cu(s) \rightarrow Cu^{2+}(aq) + 2e^-$
       $H_2SO_4(aq) + 2H^+(aq) + 2e^- \rightarrow SO_2(g) + 2H_2O(\ell)$
       $Cu(s) + H_2SO_4(aq) + 2H^+(aq) \rightarrow Cu^{2+}(aq) + SO_2(g) + 2H_2O(\ell)$

The molecular equation is:

$Cu(s) + 2H_2SO_4(aq) \rightarrow CuSO_4(aq) + SO_2(g) + 2H_2O(\ell)$

9.24   The first reaction demonstrates that Al is more readily oxidized than Cu. The second reaction demonstrates that Al is more readily oxidized than Fe. Reaction 3 demonstrates that Fe is more readily oxidized than Pb. Reaction 4 demonstrates that Fe is more readily oxidized than Cu. The fifth reaction demonstrates that Al is more readily oxidized than Pb. The last reaction demonstrates that Pb is more readily oxidized than Cu.

Altogether, the above facts constitute the following trend of increasing ease of oxidation:
Cu < Pb < Fe < Al

9.26   Any metal that is lower than hydrogen in the activity series shown in Table 9.3 of the text will react with H^+: (c) zinc and (d) magnesium.

9.28   In each case, the reaction should proceed to give the less reactive of the two metals, together with the ion of the more reactive of the two metals. The reactivity is taken from the reactivity series, Table 9.3.
(a)   N.R.
(b) $2Cr(s) + 3Pb^{2+}(aq) \rightarrow 2Cr^{3+}(aq) + 3Pb(s)$
(c) $2Ag^+(aq) + Fe(s) \rightarrow 2Ag(s) + Fe^{2+}(aq)$
(d) $3Ag(s) + Au^{3+}(aq) \rightarrow Au(s) + 3Ag^+(aq)$

9.30   In each case, the reaction should proceed to give the less reactive of the two metals, together with the ion of the more reactive of the two metals. The reactivity is taken from the reactivity series, Table 9.3.
(a) $Zn(s) + Sn^{2+}(aq) \rightarrow Zn^{2+}(aq) + Sn(s)$
(b) $2Cr(s) + 6H^+(aq) \rightarrow 2Cr^{3+}(aq) + 3H_2(g)$
(c)   N.R.
(d) $Mn(s) + Pb^{2+}(aq) \rightarrow Mn^{2+}(aq) + Pb(s)$
(e) $Zn(s) + Co^{2+}(aq) \rightarrow Zn^{2+}(aq) + Co(s)$

9.36   calcium > iron > silver > iridium

9.40   (a) $2C_6H_6(\ell) + 15O_2(g) \rightarrow 12CO_2(g) + 6H_2O(g)$
(b) $C_3H_8(g) + 5O_2(g) \rightarrow 3CO_2(g) + 4H_2O(g)$
(c) $C_{21}H_{44}(s) + 32O_2(g) \rightarrow 21CO_2(g) + 22H_2O(g)$
(d) $2C_{12}H_{26}(\ell) + 37O_2(g) \rightarrow 24CO_2(g) + 26H_2O(g)$
(e) $C_{18}H_{36}(\ell) + 27O_2(g) \rightarrow 18CO_2(g) + 18H_2O(g)$

9.42 $2CH_3OH(\ell) + 3O_2(g) \rightarrow 2CO_2(g) + 4H_2O(g)$

9.44 $C_{12}H_{22}O_{11}(s) + 12O_2(g) \rightarrow 12CO_2(g) + 11H_2O(g)$

9.46   The products will be CO_2, H_2O, and SO_2:
        $2C_2H_6S(\ell) + 9O_2(g) \rightarrow 4CO_2(g) + 2SO_2(g) + 6H_2O(g)$

9.48   (a) $2Zn(s) + O_2(g) \rightarrow 2ZnO(s)$
        (b) $4Al(s) + 3O_2(g) \rightarrow 2Al_2O_3(s)$
        (c) $2Mg(s) + O_2(g) \rightarrow 2MgO(s)$
        (d) $2Fe(s) + O_2(g) \rightarrow 2FeO(s)$
               alternatively we have: $4Fe(s) + 3O_2(g) \rightarrow 2Fe_2O_3(s)$
        (e) $2Ca(s) + O_2(g) \rightarrow 2CaO(s)$

9.50   (a) F_2 (b) P_4 (c) Br_2
        (d) S_8 (e) Cl_2 (f) S_8

9.51

$$\text{\# moles citric acid} = (31.25 \text{ mL NaOH})\left(\frac{0.100 \text{ mol NaOH}}{1000 \text{ mL NaOH}}\right)\left(\frac{1 \text{ mol citric acid}}{3 \text{ mol NaOH}}\right)$$

$$= 1.04 \times 10^{-3} \text{ mol citric acid}$$

$$\text{molar mass} = \frac{0.200 \text{ g citric acid}}{1.04 \times 10^{-3} \text{ mol citric acid}} = 192 \text{ g / mol citric acid}$$

CHAPTER TEN
Review Exercises

10.4   Since the density of water is approximately 13 times smaller than that of mercury, a barometer constructed with water as the moveable liquid would have to be some 13 times longer than one constructed using mercury. Also, the vapor pressure of water is large enough that the closed end of the barometer may fill with sufficient water vapor so as to affect atmospheric pressure readings. In fact, the measurement of atmospheric pressure would be about 18 torr too low, due to the presence of water vapor in the closed end of the barometer.

10.6   (a)   735 mm Hg = 735 torr

           (b)   # atm = (740 torr)(1 atm / 760 torr) = 0.974 atm

           (c)   738 torr = 738 mm Hg

           (d)   # torr = $(1.45 \times 10^3 \text{ Pa})(760 \text{ torr} / 1.01325 \times 10^5 \text{ Pa}) = 10.9$ torr

10.8   (a)   # kPa = $(595 \text{ torr})(1.01325 \times 10^2 \text{ kPa} / 760 \text{ torr}) = 79.3$ kPa

           (b)   # kPa = $(160 \text{ torr})(1.01325 \times 10^2 \text{ kPa} / 760 \text{ torr}) = 21.3$ kPa

           (c)   # kPa = $(0.300 \text{ torr})(1.01325 \times 10^2 \text{ kPa} / 760 \text{ torr}) = 4.00 \times 10^{-2}$ kPa

10.9   The most perfect vacuum pump can, at most, cause a column of mercury to rise to a height of that day's atmospheric pressure, since that is the maximum height of a column of mercury that the atmosphere can support. That is to say, a vacuum pump cannot induce a mercury column to rise any farther than what is forced up by atmospheric pressure. Correspondingly, the highest a pump could cause a column of water to rise will be only that amount of water that will be supported by atmospheric pressure, that is the weight of water that is equivalent to the weight of 760 mm of mercury. Since the density of mercury is 13.6 g/cm^3 whereas that of water is only 1.00 g/cm^3, the heights of the two columns are also related by the proportion 13.6 – to – 1.00. Thus the height of an equivalent column of water would be: 760 mm Hg $\times$ 13.6 = 1.03×10^4 mm. This is next converted to a value in feet:

$$\# \text{ ft} = \left(1.03 \times 10^4 \text{ mm}\right)\left(\frac{1 \text{ cm}}{10 \text{ mm}}\right)\left(\frac{1 \text{ in.}}{2.54 \text{ cm}}\right)\left(\frac{1 \text{ ft}}{12 \text{ in.}}\right) = 33.9 \text{ ft}$$

This means that the best conceivable vacuum pump (one capable of producing a perfect vacuum on the column of water being pulled up) could cause the water in the pipe attached to the pump to rise only 33.9 ft above the surface of the water in the pit. The water at the bottom of a 35 ft pit could not be removed by use of a vacuum pump.

10.12 $P_{gas} = P_{atm} + $ (height difference)

 $= 746$ mm $+ 80.0$ mm $= 826$ mm $= 826$ torr

When the mercury level in the manometer arm nearest the bulb goes down by 4.00 cm (12.50 to 8.50) the mercury in the other arm goes up by 4.00 cm. Hence, the difference in the heights of the two arms is 8.00 cm $= 80.0$ mm.

10.17 In general the combined gas law equation is: $\dfrac{P_1 V_1}{T_1} = \dfrac{P_2 V_2}{T_2}$, and in particular, for this

problem, we have: $P_2 = \dfrac{P_1 V_1 T_2}{T_1 V_2} = \dfrac{(740 \text{ torr})(2.58 \text{ L})(348.2 \text{ K})}{(297.2 \text{ K})(2.81 \text{ L})} = 796$ torr .

10.19 In general the combined gas law equation is $\dfrac{P_1 V_1}{T_1} = \dfrac{P_2 V_2}{T_2}$, and in particular, for this

problem, we have: $V_2 = \dfrac{P_1 V_1 T_2}{T_1 P_2} = \dfrac{(745 \text{ torr})(2.68 \text{ L})(648.2 \text{ K})}{(297.2 \text{ K})(760 \text{ torr})} = 5.73$ L

10.21 In general the combined gas law equation is: $\dfrac{P_1 V_1}{T_1} = \dfrac{P_2 V_2}{T_2}$, and in particular, for this

problem, we have: $T_2 = \dfrac{P_2 V_2 T_1}{P_1 V_1} = \dfrac{(373 \text{ torr})(9.45 \text{ L})(293.2 \text{ K})}{(761 \text{ torr})(6.18 \text{ L})} = 220 \text{ K} = -53 \ ^\circ\text{C}$

10.23 In general the combined gas law equation is: $\dfrac{P_1 V_1}{T_1} = \dfrac{P_2 V_2}{T_2}$, and in particular, for this

problem, we have: $P_1 = \dfrac{P_2 V_2 T_1}{T_2 V_1} = \dfrac{(789 \text{ torr})(587 \text{ mL})(295.2 \text{ K})}{(359.2 \text{ K})(532 \text{ mL})} = 715$ torr

10.25 Since volume is to decrease, pressure must increase, and we multiply the starting pressure by a volume ratio that is larger than one. Also, since $P_1 V_1 = P_2 V_2$, we can solve for P_2:

$$P_2 = \frac{P_1 V_1}{V_2} = \frac{(755 \text{ torr})(740 \text{ mL})}{(525 \text{ mL})} = 1060 \text{ torr} \ \text{(Note: 3 sig figs)}$$

10.27 (a) Since $P_1 V_1 = P_2 V_2$, we can solve for P_2, keeping in mind that the new pressure must be lower than the original pressure, since the volume is larger after the expansion of the gas:

$$P_2 = \frac{P_1 V_1}{V_2} = \frac{(525 \text{ torr})(100 \text{ mL})}{(250 \text{ mL})} = 210 \text{ torr}$$

(b) The Combined Gas Law equation is needed here: $\dfrac{P_1 V_1}{T_1} = \dfrac{P_2 V_2}{T_2}$, rearranging

gives: $P_2 = \dfrac{P_1 V_1 T_2}{V_2 T_1} = \dfrac{(525 \text{ torr})(100 \text{ mL})(288 \text{ K})}{(250 \text{ mL})(297 \text{ K})} = 204 \text{ torr}$

10.29 First calculate the initial volume (V_1) and the final volume (V_2) of the cylinder, using the given geometrical data, noting that the radius is half the diameter (10.7/2 = 5.35 cm): V_1
$= \pi \times (5.35 \text{ cm})^2 \times 13.4 \text{ cm} = 1.20 \times 10^3 \text{ cm}^3$
$V_2 = \pi \times (5.35 \text{ cm})^2 \times (13.4 \text{ cm} - 12.7 \text{ cm}) = 60 \text{ cm}^3$

In general the combined gas law equation is: $\dfrac{P_1 V_1}{T_1} = \dfrac{P_2 V_2}{T_2}$, and in particular, for this

problem, we have: $T_2 = \dfrac{P_2 V_2 T_1}{P_1 V_1} = \dfrac{(34.0 \text{ atm})(60 \text{ cm}^3)(364 \text{ K})}{(1.00 \text{ atm})(1.20 \times 10^3 \text{ cm}^3)} = 619 \text{ K} = 346\ ^\circ\text{C}$

10.34 Since PV = nRT, then;

$$V = \frac{nRT}{P} = \frac{(1.00 \text{ mol})\left(0.0821\ \frac{\text{L atm}}{\text{mol K}}\right)(293.2 \text{ K})}{(760 \text{ torr})\left(\dfrac{1 \text{ atm}}{760 \text{ torr}}\right)} = 24.1 \text{ L}$$

10.36 Since PV = nRT, then;

$$P = \frac{nRT}{V} = \frac{(4.18 \text{ mol})\left(0.0821\ \frac{\text{L atm}}{\text{mol K}}\right)(291.2 \text{ K})}{(24.0 \text{ L})} = 4.16 \text{ atm}$$

10.38 Since PV = nRT, then;

$$P = \frac{nRT}{V} = \frac{(79.8 \text{ mol})\left(0.0821\ \frac{\text{L atm}}{\text{mol K}}\right)(297.2 \text{ K})}{(10.0 \text{ L})} = 195 \text{ atm}$$

10.40 Since PV = nRT, then;

$$n = \frac{PV}{RT} = \frac{(148 \text{ atm})(25.0 \text{ L})}{\left(0.0821\ \frac{\text{L atm}}{\text{mol K}}\right)(298.2 \text{ K})} = 151 \text{ moles}$$

10.42   Since PV = nRT, then n = PV/RT

(a)

$$n = \frac{PV}{RT} = \frac{(7\ atm)(14.5\ L)}{\left(0.0821\ \frac{L\ atm}{mol\ K}\right)(298.2\ K)} = 4\ moles\ O_2$$

g O_2 = 4 moles × 32.0 g / mol = 128 g O_2

(b)   The balanced equation is: $2C_2H_2 + 5O_2 \rightarrow 4CO_2 + 2H_2O$. So,

$$\#\ moles\ C_2H_2 = (4\ moles\ O_2)\left(\frac{2\ moles\ C_2H_2}{5\ moles\ O_2}\right) = 2\ moles\ C_2H_2$$

10.44 (a)   density $C_2H_6 = \left(\frac{30.1\ g\ C_2H_6}{1\ mol\ C_2H_6}\right)\left(\frac{1\ mol}{22.4\ L}\right) = 1.34\ g\ L^{-1}$

(b)   density $N_2 = \left(\frac{28.0\ g\ N_2}{1\ mol\ N_2}\right)\left(\frac{1\ mol}{22.4\ L}\right) = 1.25\ g\ L^{-1}$

(c)   density $Cl_2 = \left(\frac{70.9\ g\ Cl_2}{1\ mol\ Cl_2}\right)\left(\frac{1\ mol}{22.4\ L}\right) = 3.17\ g\ L^{-1}$

(d)   density $Ar = \left(\frac{39.9\ g\ Ar}{1\ mol\ Ar}\right)\left(\frac{1\ mol}{22.4\ L}\right) = 1.78\ g\ L^{-1}$

10.46 In general PV = nRT,   where n = mass ÷ formula mass. Thus
$$PV = \frac{mass}{(formula\ mass)}RT$$
and we arive at the formula for the density (mass divided by volume) of a gas:

$$D = \frac{P \times (formula\ mass)}{RT}$$
$$D = \frac{(742\ torr)\left(\frac{1\ atm}{760\ torr}\right)(32.0\ g\ /\ mol)}{\left(0.0821\ \frac{L\ atm}{mol\ K}\right)(297.2\ K)}$$
$$D = 1.28\ g\ /\ L\ for\ O_2$$

10.48 The formula for the density of a gas (See exercises 10.46 and 10.47) can be rearranged to give the formula for molecular mass, if the density is given:

$$\text{formula mass} = \frac{DRT}{P}$$

$$\text{formula mass} = \frac{(1.13 \text{ g L}^{-1})(0.0821 \frac{\text{L atm}}{\text{mol K}})(295.2 \text{ K})}{(755 \text{ torr})(\frac{1 \text{ atm}}{760 \text{ torr}})}$$

$$\text{formula mass} = 27.6 \text{ g mol}^{-1}$$

10.50 As in questions 10.48 and 10.49, we first need a value for density: D = 12.1 mg/255 mL = 12.1 g/255 L = 0.0475 g/L

$$\text{formula mass} = \frac{DRT}{P}$$

$$\text{formula mass} = \frac{(0.0475 \text{ g L}^{-1})(0.0821 \frac{\text{L atm}}{\text{mol K}})(298.2 \text{ K})}{(10.0 \text{ torr})(\frac{1 \text{ atm}}{760 \text{ torr}})}$$

$$\text{formula mass} = 88.4 \text{ g mol}^{-1}$$

10.52 The density of the gas is given by the formula:

$$D = \frac{P \times (\text{formula mass})}{RT}$$

$$D = \frac{(20.0 \text{ torr})(\frac{1 \text{ atm}}{760 \text{ torr}})(27.6 \text{ g / mol})}{(0.0821 \frac{\text{L atm}}{\text{mol K}})(298.2 \text{ K})}$$

$$D = 0.0297 \text{ g L}^{-1}$$

The mass is then: $(0.455 \text{ L})(0.0297 \text{ g L}^{-1})(1000 \text{ mg / 1 g}) = 13.5 \text{ mg}$

10.54 When gases are held at the same temperature and pressure, and dispensed in this fashion during chemical reactions, then they react in a ratio of volumes that is equal to the ratio of the coefficients (moles) in the balanced chemical equation for the given reaction. We can, therefore, directly use the stoichiometry of the balanced chemical equation to determine the combining ratio of the gas volumes:

$$\# \text{ L N}_2 = (45.0 \text{ L H}_2)\left(\frac{1 \text{ volume N}_2}{3 \text{ volume H}_2}\right) = 15.0 \text{ L N}_2$$

10.56

$$\# \text{ mol } C_3H_6 = (18.0 \text{ g } C_3H_6)\left(\frac{1 \text{ mol } C_3H_6}{42.08 \text{ g } C_3H_6}\right) = 0.428 \text{ mol } C_3H_6$$

$$\# \text{ mol } H_2 = (0.428 \text{ mol } C_3H_6)\left(\frac{1 \text{ mol } H_2}{1 \text{ mol } C_3H_6}\right) = 0.428 \text{ mol } H_2$$

$$V = \frac{nRT}{P} = \frac{(0.428 \text{ mol } H_2)\left(0.0821 \frac{L \text{ atm}}{\text{mol K}}\right)(297.2 \text{ K})}{(740 \text{ torr})\left(\frac{1 \text{ atm}}{760 \text{ torr}}\right)} = 10.7 \text{ L } H_2$$

10.58 (a)

$$\# \text{ mol } C_2H_2 = (50.0 \text{ g } CaC_2)\left(\frac{1 \text{ mol } CaC_2}{64.1 \text{ g } CaC_2}\right)\left(\frac{1 \text{ mol } C_2H_2}{1 \text{ mol } CaC_2}\right)$$

$$= 0.780 \text{ mol } C_2H_2$$

$$V = \frac{nRT}{P} = \frac{(0.780 \text{ mol } C_2H_2)\left(0.0821 \frac{L \text{ atm}}{\text{mol K}}\right)(297.2 \text{ K})}{(745 \text{ torr})\left(\frac{1 \text{ atm}}{760 \text{ torr}}\right)}$$

$$= 19.4 \text{ L } C_2H_2$$

(b)

$$n = \frac{PV}{RT} = \frac{(745 \text{ torr})\left(\frac{1 \text{ atm}}{760 \text{ torr}}\right)(1.00 \times 10^6 \text{ L})}{\left(0.0821 \frac{L \text{ atm}}{\text{mol K}}\right)(297.2 \text{ K})}$$

$$= 4.02 \times 10^4 \text{ mol } C_2H_2$$

which requires 4.02×10^4 mol CaC_2 according to the stoichiometry.

$$\# \text{ kg } CaC_2 = (4.02 \times 10^4 \text{ mol } CaC_2)\left(\frac{64.1 \text{ g } CaC_2}{1 \text{ mol } CaC_2}\right)\left(\frac{1 \text{ kg}}{1000 \text{ g}}\right) = 2.58 \times 10^3 \text{ kg } CaC_2$$

10.59 (a) First determine the % by weight C and H in the sample:
% C = 1.389 g/1.620 g × 100 = 85.74 % C
% H = 0.2314 g/1.620 g × 100 = 14.28 % H

(b) Next, determine the number of moles of C and H in a sample of the material weighing 100 g exactly, in order to make the conversion from % by weight to grams straightforward: In 100 g of the material, there are 85.74 g C and 14.28 g H:

$$85.74 \text{ g C} \div 12.01 \text{ g/mol} = 7.139 \text{ mol C}$$
$$14.28 \text{ g H} \div 1.008 \text{ g/mol} = 14.17 \text{ mol H}$$

Dividing each of these mole amounts by the smaller of the two gives the relative mole amounts of C and H in the material:

for C, $7.139 \div 7.139 = 1.000$ relative moles

for H, $14.17 \div 7.139 = 1.985$ relative moles

and the empirical formula is, therefore, CH_2.

(c) We determine the molecular weight of the material by use of the ideal gas law:

$$n = \frac{PV}{RT} = \frac{(748 \text{ torr})\left(\frac{1 \text{ atm}}{760 \text{ torr}}\right)(0.941 \text{ L})}{\left(0.0821 \frac{L \text{ atm}}{\text{mol K}}\right)(293.15 \text{ K})} = 0.0385 \text{ mol}$$

The molecular weight is given by the mass in grams (given in the problem) divided by the moles determined here: mol.wt. $= 1.620 \text{ g} \div 0.0385 \text{ mol} = 42.1 \text{ g mol}^{-1}$

Since this is equal to some whole number multiple of the formula weight of the empirical unit determined in step (b) above, namely CH_2 (f.wt. = 14.0), then it follows that:

$42.1 \text{ g mol}^{-1} = n \times 14.0 \text{ g mol}^{-1}$, $\therefore n = 3$, and the molecular formula is three times the empirical formula, namely C_3H_6.

10.60 For nitrogen we have:

$$\# \text{ kPa} = (594.70 \text{ torr})\left(\frac{101.325 \text{ kPa}}{760 \text{ torr}}\right) = 79.287 \text{ kPa}$$

For oxygen:

$$\# \text{ kPa} = (160.00 \text{ torr})\left(\frac{101.325 \text{ kPa}}{760 \text{ torr}}\right) = 21.332 \text{ kPa}$$

These two gases are normally the largest constituents in air. Other gases must account for the remaining pressure.

10.62 $P_{total} = 748 \text{ torr} = (P_{O_2} + P_{H_2O})$

$P_{H_2O} = 9.209$ torr at 10 °C, from Table 10.2.

$P_{O_2} = 748 \text{ torr} - 9.209 = 739 \text{ torr}$

10.64 $P_{total} = (P_{H_2} + P_{H_2O})$

$P_{H_2O} = 23.76$ torr at 25 °C, from Table 10.2.

$P_{H_2} = P_{total} - P_{H_2O} = 742 - 23.76 = 718$ torr

The temperature stays constant so, $P_1V_1 = P_2V_2$, and

$$V_2 = \frac{P_1V_1}{P_2} = \frac{(718 \text{ torr})(288 \text{ mL})}{(760 \text{ torr})} = 272 \text{ mL}$$

10.66 First convert the needed amount of oxygen at 760 torr to the volume that would correspond to the laboratory conditions of 746 torr: $P_1V_1 = P_2V_2$ or $V_2 = P_1V_1/P_2$

$V_2 = 275$ mL × 760 torr/746 torr = 280 mL of dry oxygen gas

The wet sample of oxygen gas will also be collected at atmospheric pressure, 746 torr. The vapor pressure of water at 15 °C is equal to 12.8 torr (from Table 10.2), and the wet sample will have the following partial pressure of oxygen, once it is collected:

$P_{O_2} = P_{total} - P_{H_2O} = 746 - 12.8 = 733$ torr of oxygen in the wet sample. Thus the wet sample of oxygen is composed of the following % oxygen:

% oxygen in the wet sample = 733/746 × 100 = 98.3 %

The question now becomes what amount of a wet sample of oxygen will contain the equivalent of 280 mL of pure oxygen, if the wet sample is only 98.3 % oxygen (and 1.7 % water). $0.983 \times V_{wet} = 280$ mL, hence $V_{wet} = 285$ mL. This means that 285 mL of a wet sample of oxygen must be collected in order to obtain as much oxygen as would be present in 280 mL of a pure sample of oxygen.

10.70 The relative rates are inversely proportional to the square roots of their molecular masses:

$$\frac{rate(NH_3)}{rate(\text{unknown gas})} = \sqrt{\frac{\text{molar mass (unknown gas)}}{\text{molar mass } (NH_3)}} = 2.93$$

$$\frac{\text{molar mass (unknown gas)}}{\text{molar mass } (NH_3)} = (2.93)^2$$

molar mass (unknown gas) = molar mass $(NH_3) \times (2.93)^2$

molar mass (unknown gas) = 17.028 g / mol × 8.58 = 146 g / mol

10.71 The relative rates are inversely proportional to the square roots of their molecular masses:

$$\frac{\text{rate(U-235)}}{\text{rate(U-238)}} = \sqrt{\frac{\text{molar mass (U-238)}}{\text{molar mass (U-235)}}} = \sqrt{\frac{352 \text{ g mol}^{-1}}{349 \text{ g mol}^{-1}}} = 1.0043$$

Meaning that the rate of effusion of the U–235 is only 1.0043 times faster than the U–238 isotope.

10.85 The temperatures must first be converted to Kelvin:

$$^\circ C = \frac{5}{9} \times (^\circ F - 32) = \frac{5}{9} \times (-50 - 32) = -46\ ^\circ C$$

$$^\circ C = \frac{5}{9} \times (^\circ F - 32) = \frac{5}{9} \times (120 - 32) = 49\ ^\circ C$$

Next, the pressure calculation is done using the following equation:

$$P_2 = \frac{P_1 T_2}{T_1} = \frac{(35 \text{ lb in.}^{-2})(322 \text{ K})}{(227 \text{ K})} = 50 \text{ lb in.}^{-2}$$

10.87

$$T_2 = \frac{P_2 T_1}{P_1} = \frac{(2.50 \text{ atm})(298.2 \text{ K})}{(1.00 \text{ atm})} = 745.5 \text{ K} = 472\ ^\circ C$$

10.89 (a) $Zn(s) + 2HCl(aq) \rightarrow H_2(g) + ZnCl_2(aq)$
     (b) Calculate the number of moles of hydrogen:

$$n = \frac{PV}{RT} = \frac{(760 \text{ torr})(\frac{1 \text{ atm}}{760 \text{ torr}})(12.0 \text{ L})}{(0.0821 \frac{L \text{ atm}}{mol \text{ K}})(293.2 \text{ K})} = 0.499 \text{ mol H}_2$$

and the number of moles of zinc:

$$\# \text{ mol Zn} = (0.499 \text{ mol H}_2)\left(\frac{1 \text{ mol Zn}}{1 \text{ mol H}_2}\right) = 0.499 \text{ mol Zn}$$

The number of grams of zinc that are needed are, therefore:

$$\# \text{ g Zn} = (0.499 \text{ mol Zn})\left(\frac{65.39 \text{ g Zn}}{1 \text{ mol Zn}}\right) = 32.6 \text{ g Zn}$$

(c)   # mol HCl $= (0.499 \text{ mol Zn}) \left(\dfrac{2 \text{ mol HCl}}{1 \text{ mol Zn}} \right) = 0.998 \text{ mol HCl}$

(d)   # mL HCl $= (0.998 \text{ mol HCl}) \left(\dfrac{1000 \text{ mL HCl}}{8.00 \text{ mol HCl}} \right) = 125 \text{ mL HCl}$

10.91 $P_{total} = 745 \text{ torr} = P_{CO_2} + P_{water}$

The vapor pressure of water at 20 °C is available in Table 10.2: 17.54 torr. Hence:
$P_{CO_2} = (745 - 18) \text{ torr} = 727 \text{ torr}$

Next, we calculate the number of moles of carbon dioxide gas that this represents:

$$n = \frac{PV}{RT} = \frac{(727 \text{ torr})\left(\frac{1 \text{ atm}}{760 \text{ torr}}\right)(0.465 \text{ L})}{\left(0.0821 \frac{\text{L atm}}{\text{mol K}}\right)(293.2 \text{ K})} = 0.0185 \text{ mol } CO_2$$

Next we determine the quantities of the reagents needed:

$$\# \text{ g } CaCO_3 = (0.0185 \text{ mol } CO_2)\left(\frac{1 \text{ mol } CaCO_3}{1 \text{ mol } CO_2}\right)\left(\frac{100.09 \text{ g } CaCO_3}{1 \text{ mol } CaCO_3}\right) = 1.85 \text{ g } CaCO_3$$

$$\# \text{ mL HCl} = (0.0185 \text{ mol } CO_2)\left(\frac{2 \text{ mol HCl}}{1 \text{ mol } CO_2}\right)\left(\frac{1000 \text{ mL HCl}}{8.00 \text{ mol HCl}}\right) = 4.63 \text{ mL HCl}$$

10.93   (a)   First determine the % by weight S and O in the sample:
% S = 1.448 g/3.620 g × 100 = 40.00 % S
% O = 2.172 g/3.620 g × 100 = 60.00 % O

(b)   Next, determine the number of moles of S and O in a sample of the material weighing 100 g exactly, in order to make the conversion from % by weight to grams straightforward: In 100 g of the material, there are 40.00 g S and 60.00 g O:
40.00 g S ÷ 32.01 g/mol = 1.250 mol S
60.00 g O ÷ 16.00 g/mol = 3.750 mol O
Dividing each of these mole amounts by the smaller of the two gives the relative mole amounts of S and O in the material: for S, 1.250 ÷ 1.250 = 1.000 relative moles, for O, 3.750 ÷ 1.250 = 3.000 relative moles, and the empirical formula is, therefore, SO_3.

(c)   We determine the formula mass of the material by use of the ideal gas law:

$$n = \frac{PV}{RT} = \frac{(750 \text{ torr})\left(\frac{1 \text{ atm}}{760 \text{ torr}}\right)(1.120 \text{ L})}{\left(0.0821 \frac{\text{L atm}}{\text{mol K}}\right)(298.2 \text{ K})} = 0.0451 \text{ mol}$$

The formula mass is given by the mass in grams (given in the problem) divided by the moles determined here: formula mass = 3.620 g ÷ 0.0451 mol = 80.3 g mol^{-1}. Since this is equal to the formula weight of the empirical unit determined in step (b) above, namely SO_3, then the molecular formula is also SO_3.

10.96   (a)   We begin by converting the dimensions of the room into cm: 40 ft × 30.48 cm/ft = 1.2×10^3 cm, 20 ft × 30.48 cm/ft = 6.1×10^2 cm, 8 ft × 30.48 cm/ft = 2.4×10^2 cm. Next, the volume of the room is determined: V = $(1.2 \times 10^3$ cm$)(6.1 \times 10^2$ cm$)(2.4 \times 10^2$ cm$) = 1.8 \times 10^8$ cm^3. Since there are 1000 cm^3 in a liter, volume is: V = 1.8×10^5 L

The calculation of the amount of H_2S goes as follows:

$$\# \text{ L H}_2\text{S} = \left(1.8 \times 10^5 \text{ L space}\right)\left(\frac{0.15 \text{ L H}_2\text{S}}{1 \times 10^9 \text{ L space}}\right) = 2.7 \times 10^{-5} \text{ L H}_2\text{S}$$

(b)   Convert volume (in liters) to moles at STP:

$$\# \text{ mol H}_2\text{S} = \left(2.7 \times 10^{-5} \text{ L H}_2\text{S}\right)\left(\frac{1 \text{ mol H}_2\text{S}}{22.4 \text{ L H}_2\text{S}}\right) = 1.2 \times 10^{-6} \text{ mol H}_2\text{S}$$

Since the stoichiometry is 1:1, we require the same number of moles of Na_2S:

$$\# \text{ mL Na}_2\text{S} = \left(1.2 \times 10^{-6} \text{ mol Na}_2\text{S}\right)\left(\frac{1000 \text{ mL Na}_2\text{S}}{0.100 \text{ mol Na}_2\text{S}}\right)$$

$$= 1.2 \times 10^{-2} \text{ mL Na}_2\text{S}$$

10.97 Using the ideal gas law, detemine the number of moles of H_2 and O_2 gas initially present:

for hydrogen:

$$n = \frac{PV}{RT} = \frac{(1250 \text{ torr})(\frac{1 \text{ atm}}{760 \text{ torr}})(400 \text{ mL})(\frac{1 \text{ L}}{1000 \text{ mL}})}{(0.0821 \frac{\text{L atm}}{\text{mol K}})(318 \text{ K})} = 2.52 \times 10^{-2} \text{ mol } H_2$$

for oxygen:

$$n = \frac{PV}{RT} = \frac{(740 \text{ torr})(\frac{1 \text{ atm}}{760 \text{ torr}})(300 \text{ mL})(\frac{1 \text{ L}}{1000 \text{ mL}})}{(0.0821 \frac{\text{L atm}}{\text{mol K}})(298 \text{ K})} = 1.19 \times 10^{-2} \text{ mol } O_2$$

This problem is an example of a limiting reactant problem in that we know the amounts of H_2 and O_2 initially present. Since 1 mol of O_2 reacts completely with 2 mol of H_2, we can see, by inspection, that there is excess H_2 present. Using the amounts calculated above, we can make 1.19×10^{-2} mol of H_2O and have an excess of 1.4×10^{-3} mol of H_2. Thus, the total amount of gas present after complete reaction is 1.33×10^{-2} mol. Using the value for n, we can calculate the final pressure in the reaction vessel:

$$P = \frac{nRT}{V} = \frac{(1.33 \times 10^{-2} \text{ mol})(0.0821 \frac{\text{L atm}}{\text{mol K}})(393 \text{ K})}{(500 \text{ mL})(\frac{1 \text{ L}}{1000 \text{ mL}})} = 0.858 \text{ atm} = 652 \text{ torr}$$

10.98 We first need to determine the pressure inside the apparatus. Since the water level is 8.5 cm (= 85 mm = 85 torr) higher inside than outside, the pressure inside the container is higher than the pressure outside: $P_{inside} = P_{outside} + 85 \text{ torr} = 764 \text{ torr} + 85 \text{ torr} = 831$ torr. In order to determine the P_{H_2}, we need to subtract the vapor pressure of water at 24 °C. This value may be found in Appendix E4 and is equal to 22.4 torr. The $P_{H_2} = P_{inside} - P_{H_2O} = 831 \text{ torr} - 22.4 \text{ torr} = 809 \text{ torr}$. Now, we can use the ideal gas law in order to determine the number of moles of H_2 present;

$$n = \frac{PV}{RT} = \frac{(809 \text{ torr})(\frac{1 \text{ atm}}{760 \text{ torr}})(18.45 \text{ mL})(\frac{1 \text{ L}}{1000 \text{ mL}})}{(0.0821 \frac{\text{L atm}}{\text{mol K}})(297 \text{ K})} = 8.05 \times 10^{-4} \text{ mol}$$

The balanced equation described in this problem is:

$$Zn(s) + 2HCl(aq) \rightarrow ZnCl_2(aq) + H_2(g)$$

By inspection we can see that 1 mole of $Zn(s)$ reacts to form 1 mole of $H_2(g)$ and we must have reacted 8.05×10^{-4} mol Zn in this reaction.

$$\# \, g \, Zn = \left(8.05 \times 10^{-4} \, \text{mol Zn}\right)\left(\frac{65.39 \, g \, Zn}{1 \, \text{mol Zn}}\right) = 5.3 \times 10^{-2} \, g \, Zn$$

11.9 Since these are both nonpolar molecular substances, the only type of intermolecular force that we need to consider is London forces. The larger molecule has the greater London force of attraction and hence the higher boiling point: C_8H_{18}.

11.14 London forces are possible in them all. Where another intermolecular force can operate, it is generally stronger than London forces, and this other type of interaction overshadows the importance of the London force. The substances in the list that can have dipole–dipole attractions are those with permanent dipole moments: (a), (c), and (e). Both CS_2 and SF_6 are nonpolar molecular substances. HF, (a), has hydrogen bonding.

11.23 Water should have the greater surface tension because it has the stronger intermolecular force, i.e., hydrogen bonding.

11.26 Glycerol ought to wet the surface of glass quite nicely, because the oxygen atoms at the surface of glass can form effective hydrogen bonds to the O–H groups of glycerol.

11.30 Diethyl ether has the faster rate of evaporization, since it does not have hydrogen bonds, as does butanol.

11.31 The snow dissipates by sublimation.

11.43 Diethyl ether, which lacks hydrogen bonds, should have the higher vapor pressure.

11.45 Air with 100% humidity is saturated with water vapor, meaning that the partial pressure of water in the air has become equal to the vapor pressure of water at that temperature. Since vapor pressure increases with increasing temperature, so too should the total amount of water vapor in humid (saturated) air increase with increasing temperature.

11.47 At the temperature of the cool glass, the equilibrium vapor pressure of the water is lower than the partial pressure of water in the air. The air in contact with the cool glass is induced to relinquish some of its water, and condensation occurs.

11.49 Although the cold air outside the building may be nearly saturated at the low temperature, the same air inside at a higher temperature is not now saturated. In fact the water content of the air at the higher temperature may be only a fraction of the maximum % humidity for that temperature. The % humidity of the air at the indoor temperature is, therefore, comparatively low.

11.53 at about 73 °C

11.55 Butanol should have the higher boiling point, since it has strong hydrogen bonds.

11.59 Inside the lighter, the liquid butane is in equilibrium with its vapor, which exerts a pressure somewhat above normal atmospheric pressure. If the liquid butane were spilled on a desk top, it would vaporize (boil) because it would be at a pressure of about 1 atm only, which is somewhat less than the vapor pressure of butane. The intermolecular forces in butane are weak.

11.61 The hydrogen bond network in HF is less extensive than in water, because it is a monohydride not a dihydride.

11.64 $\text{\# kJ} = (125 \text{ g H}_2\text{O})\left(\dfrac{1 \text{ mol H}_2\text{O}}{18.015 \text{ g H}_2\text{O}}\right)\left(\dfrac{43.9 \text{ kJ}}{1 \text{ mol H}_2\text{O}}\right) = 305 \text{ kJ}$

11.68 Water should have the higher heat of vaporization; its boiling point is higher and, hence, its intermolecular forces are stronger.

11.70 The substance with the larger heat of vaporization has the stronger intermolecular forces. This is ethanol, which has hydrogen bonding, whereas ethyl acetate does not.

11.72 We can approach this problem by first asking either of two equivalent questions about the system: how much heat energy (q) is needed in order to melt the entire sample of solid water (105 g), or how much energy is lost when the liquid water (45.0 g) is cooled to the freezing point? Regardless, there is only one final temperature for the combined (150.0 g) sample, and we need to know if this temperature is at the melting point (0 °C, at which temperature some solid water remains in equilibrium with a certain amount of liquid water) or above the melting point (at which temperature all of the solid water will have melted).

Heat flow supposing that all of the solid water is melted:
q = 6.01 kJ/mole × 105 g × 1 mol/18.0 g = 35.1 kJ

Heat flow on cooling the liquid water to the freezing point:
q = 45.0 g × 4.18 J/g °C × 85 °C = 1.60×10^{4} J = 16.0 kJ

The lesser of these two values is the correct one, and we conclude that 16.0 kJ of heat energy will be transferred from the liquid to the solid, and that the final temperature of the mixture will be 0 °C. The system will be an equilibrium mixture weighing 150 g and having some solid and some liquid in equilibrium with one another. The amount of solid that must melt in order to decrease the temperature of 45.0 g of water from 85 °C to 0 °C is: 16.0 kJ ÷ 6.01 kJ/mol = 2.66 mol of solid water. 2.66 mol × 18.0 g/mol = 47.9 g of water must melt.

(a)   The final temperature will be 0 °C.
(b)   47.9 g of water must melt.

11.74 $CH_4 < CF_4 < HCl < HF$

11.76

$$\ln\left(\frac{P_1}{P_2}\right) = \frac{\Delta H_{vap}}{8.314 \text{ J mol}^{-1}\text{K}^{-1}}\left(\frac{1}{T_2} - \frac{1}{T_1}\right)$$

$$\ln\left(\frac{31.6}{P_2}\right) = \frac{42.09 \times 10^3 \text{ J mol}^{-1}}{8.314 \text{ J mol}^{-1}\text{K}^{-1}}\left(\frac{1}{333 \text{ K}} - \frac{1}{293 \text{ K}}\right)$$

Take the antilog of both sides

$$\left(\frac{31.6}{P_2}\right) = e^{-2.08} = 0.126$$

$P_2 = 31.6 \text{ torr} / 0.126 = 252 \text{ torr}$

11.79 This is an endothermic system, and adding heat to the system will shift the position of the equilibrium to the right, producing a new equilibrium mixture having more liquid and less solid. Some of the solid melts when heat is added to the system.

11.86 Carbon dioxide does not have a normal boiling point because its triple point lies above one atmosphere. Thus, the liquid–vapor equilibrium that is taken to represent the boiling point does not exist at the pressure (1 atm) conventionally used to designate the "normal" boiling point.

11.88 The solid's temperature increases until the solid–liquid line is reached, at which point the solid sublimes. After it has completely vaporized, the vapor's temperature increases to 0 °C.

11.92 The critical temperature of hydrogen is below room temperature because, at room temperature, it cannot be liquified by the application of pressure. The critical temperature of butane is above room temperature, because butane can be liquified by the application of pressure. See also the answer to Review Exercise 11.59.

11.101 Each atom of a cubic unit cell is shared by eight unit cells. This means that only one eighth of each corner atom can be assigned to a given unit cell. Eight corner atoms times 1/8 each assigned to a given unit cell yields one atom per unit cell.

11.102 A cube has six faces and eight corners. Each of the six face atoms is shared by two adjacent unit cells: $6 \times 1/2 = 3$ atoms. The eight corner atoms are each shared by eight unit cells: $8 \times 1/8 = 1$ atom. The total number atoms to be assigned to any one cell is thus $3 + 1 = 4$.

11.104 The following diagram is appropriate:

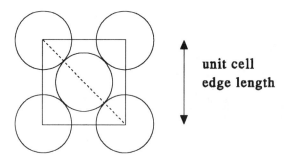

The face diagonal is 4 times the radius of the atom. The Pythagorean theorem is: $\text{diagonal}^2 = \text{edge}^2 + \text{edge}^2$. Hence we have: $[4(144 \text{ pm})]^2 = 2 \times \text{edge}^2$. Solving for the edge length we get 407 pm.

11.109 This must be a molecular solid, because if it were ionic it would be high–melting, and the melt would conduct.

11.111 This is a covalent solid.

11.113 This is a covalent, molecular solid.

11.114 This is metallic.

11.120 The principal attractive forces are ion–ion forces and ion–induced dipole attractions. These are of overwhelming strength compared to London forces, which do technically exist.

11.122 The comparatively weak intermolecular forces in acetone allow for very rapid evaporization. This provides for significantly more cooling than does ethylene glycol. The intermolecular forces of ethylene glycol, plus its ability to adhere to the skin through OH groups, causes it to evaporate slowly.

11.124 At the higher temperature, all of the carbon dioxide exists in the gas phase, because the critical temperature has been exceeded.

11.126 Here we need only solve for the temperature T_2 at which the vapor pressure P_2 has reached 760 torr:

$$\ln\left(\frac{P_1}{P_2}\right) = \frac{\Delta H_{vap}}{8.314 \text{ J mol}^{-1}\text{K}^{-1}}\left(\frac{1}{T_2} - \frac{1}{T_1}\right)$$

$$\ln\left(\frac{31.6}{760}\right) = \frac{42.09 \text{ X } 10^3 \text{ J / mol}}{8.314 \text{ J mol}^{-1}\text{K}^{-1}}\left(\frac{1}{T_2} - \frac{1}{293 \text{ K}}\right)$$

$$-3.18 = 5063 \text{ K} \times \left(\frac{1}{T_2} - \frac{1}{293 \text{ K}}\right)$$

$$-6.28 \text{ X } 10^{-4} \text{ K}^{-1} = \frac{1}{T_2} - \frac{1}{293 \text{ K}}$$

$$\frac{1}{T_2} = 2.78 \text{ X } 10^{-3} \text{ K}^{-1}; \quad T_2 = 359 \text{ K} = 86\,^{\circ}\text{C}$$

11.128 Each edge is composed of 2 × radius of the cation plus 2 × radius of the anion. The edge is therefore 2 × 133 + 2 × 195 = 656 pm.

12.17 This is to be very much like that shown in Figures 12.7 and 12.8:

   (a) $KCl(s) \rightarrow K^+(g) + Cl^-(g)$, $\Delta H = +690 \text{ kJ mol}^{-1}$

          $K^+(g) + Cl^-(g) \rightarrow K^+(aq) + Cl^-(aq)$, $\Delta H = -686 \text{ kJ mol}^{-1}$

   (b) $KCl(s) \rightarrow K^+(aq) + Cl^-(aq)$, $\Delta H = +4 \text{ kJ mol}^{-1}$

12.18 Two percent of 690 kJ/mol is $0.02 \times 690 = 13.8$ kJ, and a value that is 2 % greater than 690 is $690 + 13.8 = 704$ kJ/mol. If this value is used in the calculation of Review Exercise 12.17, the net ΔH is now $704 - 686 = +18$ kJ/mol. A small difference (only 2 % in this case) in one part of the process leads to a large difference (450 %) in the overall value for the process.

12.24 We can estimate from Figure 12.13 that the solubility of NH_4NO_3 in 100 g of H_2O is 500 g at 70 °C and 150 g at 10 °C. The amount of solid that will crystallize is the difference between these two solubilities, namely $500 - 150 = 350$ g.

12.27 Henry's Law is the statement, applied to the dissolving of a gas in a solvent, that at a given temperature, the concentration (C_g) of the gas in a solution is directly proportional to the partial pressure (P_g) of the gas on the solution, where k in the following equation is the constant of proportionality: $C_g = k \times P_g$. As discussed in the text, an alternate statement expresses the relationship of concentration at one pressure P_1 to the concentration that would exist at some new pressure P_2: $C_1/P_1 = C_2/P_2$. Using the alternate statement of Henry's Law: $C_1/P_1 = C_2/P_2$, $\therefore$ $C_2 = (C_1 \times P_2)/P_1 = (0.025 \text{ g/L} \times 1.5 \text{ atm})/1.0 \text{ atm} = 0.038$ g/L.

12.32 11.5 g NaCl $\div$ 58.5 g/mol = 0.197 mol NaCl
molality = 0.197 mol NaCl/1.00 kg H_2O = 0.197 molal
Since the density of water is 1.00 g/mL, the volume of 1 kg is 1 L. Thus the molarity is: 0.197 mol/1.00 L = 0.197 M. A solvent must have a density close to 1 g/mL for this to happen. Also the volume of the solvent must not change appreciably on addition of the solute.

12.34 (a) If we choose, for convenience, an amount of solution that contains 1 kg of solvent, then it also contains 0.363 moles of $NaNO_3$. The number of moles of solvent is:

$$1.00 \times 10^3 \text{ g} \div 18.0 \text{ g/mol} = 55.6 \text{ mol } H_2O$$

Now, convert the number of moles to a number of grams: for $NaNO_3$, 0.363 mol × 85.0 g/mol = 30.9 g; for H_2O, 1000 g was assumed and the percent (w/w) values are: % $NaNO_3$ = 30.9 g/1031 g × 100 = 3.00 %, % H_2O = 1000 g/1031 g × 100 = 97.0 %.

(b) First determine the mass of 1.00 L of this solution, using the known density: 1000 mL × 1.0185 g/mL = 1018.5 g. Next, we use the fact that 3.00 % of any sample of this solution is $NaNO_3$, and calculate the mass of $NaNO_3$ that is contained in 1000 mL of the solution: 0.0300 × 1018.5 g = 30.6 g $NaNO_3$. The number of moles of $NaNO_3$ is given by: 30.6 g ÷ 85.0 g/mol = 0.359 mol $NaNO_3$. The molarity is the number of moles of $NaNO_3$ per liter of solution: 0.359 mol/1.00 L = 0.359 M

12.37 (a) The mass of one L of this solution is: 1000 mL × 0.9787 g/mL = 978.7 g. The mass of NH_3 in this much solution is: 978.7 × 0.0500 = 48.94 g NH_3

$$48.94 \text{ g } NH_3 \div 17.04 \text{ g/mol} = 2.872 \text{ mol } NH_3$$

molarity = 2.872 mol/1.000 L = 2.872 M NH_3

(b) The mass of water in one L of this solution is: (978.7 g – 48.94 g) × 1 kg/1000 g = 0.9298 kg H_2O

molality = 2.872 mol NH_3/0.9298 kg H_2O = 3.089 m

12.42 $P_{solution} = P°_{solvent} \times X_{solvent}$

We need to determine $X_{solvent}$. In Review Exercise 12.31 we determined there are 0.133 moles of glucose. The number of moles of solvent is:

$$\text{\# mol } H_2O = (1000 \text{ g } H_2O)\left(\frac{1 \text{ mol } H_2O}{18.02 \text{ g } H_2O}\right) = 55.49 \text{ mol } H_2O$$

The total number of moles is thus: 55.49 mol + 0.133 mol = 55.63 mol and the mole fraction of the solvent is: $X_{solvent} = \left(\dfrac{55.49 \text{ mol solvent}}{55.63 \text{ mol solution}}\right) = 0.997$. Therefore,

$P_{solution}$ = 23.8 torr × 0.997 = 23.7 torr.

12.44 The mole fraction values are:

$X_{pentane} = 25.0\,\%/100 = 0.250$ and $X_{heptane} = 75.0\,\%/100 = 0.750$

$P_{pentane} = X_{pentane} \times P°_{pentane} = 0.250 \times 420$ torr $= 105$ torr

$P_{heptane} = X_{heptane} \times P°_{heptane} = 0.750 \times 36$ torr $= 27$ torr

$P_{Total} = P_{pentane} + P_{heptane} = (105 + 27)$ torr $= 132$ torr $= 130$ torr (2 significant figures)

12.46 (a) $P_{solvent} = X_{solvent} \times P°_{solvent}$

        336.0 torr $= X_{solvent} \times 400.0$ torr $\therefore X_{solvent} = 0.8400$

        $X_{solute} = 1 - 0.8400 = 0.1600$

(b)   The number of moles of solvent is: 33.25 g $\div 109.0$ g/mol $= 0.3050$ mol and the following expression for mole fraction of solvent can be solved to determine the number of moles of solute:

$$0.8400 = \frac{0.3050}{\text{mol solute} + 0.3050} \qquad \therefore \text{ mol solute} = 0.0581$$

(c) 18.26 g/0.0581 mol $= 314$ g/mol

12.50 (a)   Since -40 °F is also equal to -40 °C, the following expression applies: $\Delta T = K_f m$,

so 40 °C $= (1.86$ °C kg mol$^{-1}) \times m, \therefore m = 40/1.86$ mol/kg $= 22$ molal

Therefore, 22 moles must be added to 1 kg of water.

(b)

$$\# \text{ mL} = (22 \text{ moles})\left(\frac{62.1 \text{ g}}{1 \text{ mol}}\right)\left(\frac{1.00 \text{ mL}}{1.11 \text{ g}}\right) = 1.2 \times 10^3 \text{ mL}$$

(c)   There are 946 mL in one quart. Thus for 1 qt of water we are to have 946 mL, and the required number of quarts of ethylene glycol is:

$$\frac{\# \text{ qt C}_2\text{H}_6\text{O}_2}{1 \text{ qt H}_2\text{O}} = \left(\frac{1.2 \times 10^3 \text{ mL C}_2\text{H}_6\text{O}_2}{1000 \text{ g H}_2\text{O}}\right)\left(\frac{1 \text{ g H}_2\text{O}}{1 \text{ mL H}_2\text{O}}\right)\left(\frac{946 \text{ mL H}_2\text{O}}{1 \text{ qt H}_2\text{O}}\right)\left(\frac{1 \text{ qt C}_2\text{H}_6\text{O}_2}{946 \text{ mL C}_2\text{H}_6\text{O}_2}\right)$$

$$= 1.2 \text{ qt C}_2\text{H}_6\text{O}_2$$

The proper ratio of ethylene glycol to water is 1.2 qt to 1 qt.

12.52 The number of moles of glycerol is: 46.0 g $\div 92$ g/mol $= 0.50$ mol and the molality of this solution is 0.50 mol/0.250 kg $= 2.0$ m.

(a) $\Delta T = K_b \times m = 0.51$ °C kg mol$^{-1} \times 2.0$ mol kg$^{-1} = 1.0$ °C, and the boiling point is $100.0 + 1.0 = 101$ °C

(b) $\Delta T = K_f \times m = 1.86$ °C kg mol$^{-1} \times 2.0$ mol kg$^{-1} = 3.72$ °C, and the freezing point is $0.0 - 3.72 = -3.72$ °C

(c) $P_{solution} = P°_{solvent} \times X_{solvent}$. Now the number of moles of solute is: 46.0 g ÷ 92.1 g/mol = 0.499 mol $C_3H_8O_3$. The number of moles of solvent is: 250 g H_2O ÷ 18.02 g/mol = 13.9 mol H_2O, and the mole fraction of solvent is: X_{H_2O} = 13.9/(13.9 + 0.499) = 0.965. Thus, $P_{solution}$ = 23.8 torr × 0.965 = 23.0 torr

12.54 $\Delta T = K_b \times m = (81.7 - 80.2)$, ∴ m = 0.59 mol solute/kg benzene. 0.59 mol/kg benzene × 1.0 kg benzene = 0.59 mol solute and the formula weight is: 14 g/0.59 mol = 24 g/mol

12.56 (a) The formula weights are $Na_2Cr_2O_7 \cdot 2H_2O$: 298 g/mol, C_3H_8O: 60.1 g/mol, and C_3H_6O: 58.1 g/mol.

$$\# \ g \ Na_2Cr_2O_7 \bullet 2H_2O \ = \ (21.4 \ g \ C_3H_8O)\left(\frac{1 \ mol \ C_3H_8O}{60.1 \ g \ C_3H_8O}\right)$$

$$\times \left(\frac{1 \ mol \ Na_2Cr_2O_7 \bullet 2H_2O}{3 \ mol \ C_3H_8O}\right)\left(\frac{298 \ g \ Na_2Cr_2O_7 \bullet 2H_2O}{1 \ mol \ Na_2Cr_2O_7 \bullet 2H_2O}\right)$$

$$= \ 35.4 \ g \ Na_2Cr_2O_7 \bullet 2H_2O$$

(b) The theoretical yield is:

$$\# \ g \ C_3H_6O \ = \ (21.4 \ g \ C_3H_8O)\left(\frac{1 \ mol \ C_3H_8O}{60.1 \ g \ C_3H_8O}\right)\left(\frac{3 \ mol \ C_3H_6O}{3 \ mol \ C_3H_8O}\right)\left(\frac{58.1 \ g \ C_3H_6O}{1 \ mol \ C_3H_6O}\right)$$

$$= \ 20.7 \ g \ C_3H_6O$$

The percent yield is therefore: 12.4/20.7 × 100 = 59.9 %

(c) First, we determine the number of grams of C, H, and O that are found in the products, and then the % by weight of C, H, and O that were present in the sample that was analyzed by combustion, i.e. the by–product:

$$\# \ g \ C \ = \ (22.368 \ X \ 10^{-3} \ g \ CO_2)\left(\frac{12.011 \ g \ C}{44.010 \ g \ CO_2}\right) \ = \ 6.1046 \ X \ 10^{-3} \ g \ C$$

and the % C is: 6.1046 X 10^{-3} g/8.654 X 10^{-3} g × 100 = 70.54 % C

$$\# \ g \ H \ = \ (10.655 \ X \ 10^{-3} \ g \ H_2O)\left(\frac{2.0159 \ g \ H}{18.015 \ g \ H_2O}\right) \ = \ 1.1923 \ X \ 10^{-3} \ g \ H$$

and the % H is: 1.1923 X 10^{-3} g H/8.654 X 10^{-3} g × 100 = 13.78 % H

For O, the mass is the total mass minus that of C and H in the sample that was analyzed: 8.654×10^{-3} g total $- (6.1046 \times 10^{-3}$ g C $+ 1.1923 \times 10^{-3}$ g H$) = 1.357 \times 10^{-3}$ g O, and the % O is: 1.357×10^{-3} g/8.654×10^{-3} g $\times 100 = 15.68$ % O. Alternatively, we could have determined the amount of oxygen by using the mass % values, realizing that the sum of the weight percent values should be 100. Next, we convert these mass amounts for C, H, and O into mole amounts by dividing the amount of each element by the atomic weight of each element:

For C, 6.1046×10^{-3} g C $\div 12.011$ g/mol $= 0.50825 \times 10^{-3}$ mol C

For H, 1.1923×10^{-3} g H $\div 1.0079$ g/mol $= 1.1829 \times 10^{-3}$ mol H

For O, 1.357×10^{-3} g O $\div 16.00$ g/mol $= 0.08481 \times 10^{-3}$ mol O

Lastly, these are converted to relative mole amounts by dividing each of the above mole amounts by the smallest of the three (We can ignore the 10^{-3} term since it is common to all three components):

For C, $0.50825/0.08481 = 5.993$

For H, $1.1829/0.08481 = 13.95$

For O, $0.08481/0.08481 = 1.000$

and the empirical formula is given by this ratio of relative mole amounts, namely $C_6H_{14}O$.

(d) $\Delta T_f = K_f m$, $(5.45\ ^\circ C - 4.87\ ^\circ C) = (5.07\ ^\circ C/m) \times m$, $\therefore m = 0.11$ molal, and there are 0.11 moles of solute dissolved in each kg of solvent. Thus, the number of moles of solute that have been used here is: 0.11 mol/kg $\times 0.1150$ kg $= 1.31 \times 10^{-2}$ mol solute. The formula weight is thus: 1.338 g/0.0131 mol $= 102$ g/mol. Since the empirical formula has this same mass, we conclude that the molecular formula is the same as the empirical formula, i.e. $C_6H_{14}O$.

12.63   In each case, the osmotic pressure Π is given by the equation: $\Pi = M \times R \times T$.. Since we do not know either the density of the solution or the volume of the solution, we cannot convert values for % by weight into molarities. However, we do know that glucose, having the smaller molecular weight, has the higher molarity, and we conclude that it will have the larger osmotic pressure.

12.68   The dissociation of sodium chloride is represented as follows:

$NaCl(s) \rightarrow Na^+(aq) + Cl^-(aq)$ and the molality of dissolved and dissociated solute particles is: 0.171 mol NaCl/kg solvent $\times 2$ mol particles/mol NaCl $= 0.342$ molal. The number of moles of solvent in 1000 g (1 kg) is: 1000 g $H_2O \div 18.02$ g/mol $= 55.6$ mol H_2O. The solvent mole fraction is thus: $X_{H_2O} = 55.6/(55.6 + 0.342) = 0.994$, and the vapor pressure is: $P_{solution} = P^\circ_{H_2O} \times X_{H_2O} = 17.5$ torr $\times 0.994 = 17.4$ torr.

12.70   The solute that dissolves to produce the greater number of ions, Na_2CO_3, gives the solution with the larger boiling point elevation.

12.74   (a) $\Delta T_f = i \times K_f \times m = 1.89 \times 1.86\ °C/m \times 0.118\ m = 0.415\ °C$, $\therefore T_f = -0.415\ °C$

         (b)   This is due to the smaller size of the lithium cation, which is strongly hydrated.

12.76 $\Delta T_f = 0.261\ °C = K_f \times$ (apparent molality). Thus the apparent molality of solute particles

is: $m = 0.261\ °C/(1.86\ °C\ molal^{-1}) = 0.140$ molal. If the solute were dissolved as a nonelectrolyte, the apparent molality would be 0.125. The excess apparent molality arises from dissociation of the solute, and the amount $(0.140 - 0.125) = 0.015$ is the excess molality due to dissociation in this case. That is, 0.015 mol of solute per kg of solvent have been generated by dissociation of some certain % of the solute.

$$\% \text{ ionization} = 0.015/0.125 \times 100 = 12\ \%$$

12.88   (a)   Since the molarity of the solution is 4.613 mol/L, then one L of this solution contains: $4.613\ mol \times 46.08\ g/mol = 212.6\ g\ C_2H_5OH$. The mass of the total 1 L of solution is: $1000\ mL \times 0.9677\ g/mL = 967.7\ g$. The mass of water is thus $967.7\ g - 212.6\ g = 755.1\ g\ H_2O$, and the molality is: $4.613\ mol\ C_2H_5OH/0.7551$ $kg\ H_2O = 6.109\ m$.

         (b) $\%\ (w/w)\ C_2H_5OH = 212.6\ g/967.7\ g \times 100 = 21.97\ \%$

         (c)   Vol of ethyl alcohol $= 212.6\ g \div 0.7893\ g/mL = 269.4\ mL$
                Vol of $H_2O = 755.1\ g \div 0.9982\ g/mL = 756.5\ mL$
                $\%\ (v/v)\ C_2H_5OH = 269.4\ mL/(269.4\ mL + 756.5\ mL) \times 100 = 26.26\ \%$

13.8 $\Delta E = q + w = 300\ J + 700\ J = +1000\ J$

The overall process is endothermic, meaning that the internal energy of the system increases. Notice that both terms, q and w, contribute to the increase in internal energy of the system; the system gains heat (+q) and has work done on it (+w).

13.14   work $= P \times \Delta V$
The total pressure is atmospheric pressure plus that caused by the hand pump:
$P = (30.0 + 14.7)\ lb/in^2 = 44.7\ lb/in^2$

Converting to atmospheres we get:
$P = 44.7\ lb/in^2 \times 1\ atm/14.7\ lb/in^2 = 3.04\ atm$

Next we convert the volume change in units in^3 to units L:
$24.0\ in^3 \times (2.54\ cm/in)^3 \times 1\ L/1000\ cm^3 = 0.393\ L$

Hence $P \times \Delta V = (3.04\ atm)(0.393\ L) = 1.19\ L{\cdot}atm$
$1.19\ L{\cdot}atm \times 101.3\ J/L{\cdot}atm = 121\ J$

— 13.18   This is a reaction that produces 1 mol of a single gaseous product, CO_2. Furthermore, this mole of gaseous product forms from nongaseous materials. The volume that will be occupied by this gas, once it is formed, can be found by application of Charles' Law: 22.4 $L \times 298\ K/273\ K = 24.5\ L$

The work of gas expansion on forming the products is thus:
$P \times \Delta V = (1.00\ atm)(24.5\ L) = 24.5\ L\ atm$
$24.5\ l\ atm \times 101\ J/L\ atm = 2.47 \times 10^3\ J$ of work

13.19   We use the data supplied in Appendix E.
(a) $3PbO(s) + 2NH_3(g) \rightarrow 3Pb(s) + N_2(g) + 3H_2O(g)$

$\Delta H° = \{3\Delta H_f°[Pb(s)] + \Delta H_f°[N_2(g)] + 3\Delta H_f°[H_2O(g)]\}$
$\qquad - \{3\Delta H_f°[PbO(s)] + 2\Delta H_f°NH_3(g)]\}$
$\Delta H° = \{3\ mol \times (0\ kJ/mol) + 1\ mol \times (0\ kJ/mol) + 3\ mol \times (-242\ kJ/mol)\}$
$\qquad - \{3\ mol \times (-217.3\ kJ/mol) + 2\ mol \times (-46.0\ kJ/mol)\}$
$\Delta H° = +18\ kJ$

$\Delta E = \Delta H° - \Delta nRT$

$\Delta E = 18 \text{ kJ} - (+2 \text{ mol})(8.314 \text{ J/mol K})(10^{-3} \text{ kJ/J})(298 \text{ K}) = 13 \text{ kJ}$

(b) $NaOH(s) + HCl(g) \rightarrow NaCl(s) + H_2O(\ell)$

$\Delta H° = \{\Delta H_f°[NaCl(s)] + \Delta H_f°[H_2O(\ell)]\} - \{\Delta H_f°[NaOH(s)] + \Delta H_f°[HCl(g)]\}$
$\Delta H° = \{1 \text{ mol} \times (-413 \text{ kJ/mol}) + 1 \text{ mol} \times (-286 \text{ kJ/mol})\}$
$\qquad\qquad - \{1 \text{ mol} \times (-426.8 \text{ kJ/mol}) + 1 \text{ mol} \times (-92.5)\}$
$\Delta H° = -180 \text{ kJ}$

$\Delta E = \Delta H° - \Delta nRT$
$\Delta E = -180 \text{ kJ} - (-1)(8.314 \text{ J/mol K})(10^{-3} \text{ kJ/J})(298 \text{ K}) = -178 \text{ kJ}$

(c) $C_2H_4(g) + H_2O(g) \rightarrow C_2H_5OH(\ell)$

$\Delta H° = \{\Delta H_f°[C_2H_5OH(\ell)] - \{\Delta H_f°[C_2H_4(g)] + \Delta H_f°[H_2O(g)]\}$
$\Delta H° = \{1 \text{ mol} \times (-278 \text{ kJ/mol}) \}$
$\qquad\qquad - \{1 \text{ mol} \times (+51.9 \text{ kJ/mol}) + 1 \text{ mol} \times (-242 \text{ kJ/mol})\}$
$\Delta H° = -88 \text{ kJ}$

$\Delta E = \Delta H° - \Delta nRT$
$\Delta E = -88 \text{ kJ} - (-2)(8.314 \text{ J/mol K})(10^{-3} \text{ kJ/J})(298 \text{ K}) = -83 \text{ kJ}$

(d) $2CH_4(g) \rightarrow C_2H_6(g) + H_2(g)$
$\Delta H° = \{\Delta H_f°[C_2H_6(g)] + \Delta H_f°[H_2(g)]\} - \{2\Delta H_f°[CH_4(g)]\}$
$\Delta H° = \{1 \text{ mol} \times (-84.5 \text{ kJ/mol}) + 1 \text{ mol} \times (0.0 \text{ kJ/mol})\}$
$\qquad\qquad - \{2 \text{ mol} \times (-74.9 \text{ kJ/mol})\}$
$H° = 65.3 \text{ kJ}$

$\Delta E = \Delta H°$, since the value of Δn for this reaction is zero.

13.25 In general, we have the equation: $\Delta H° = (\text{sum } \Delta H_f°[\text{products}]) - (\text{sum } \Delta H_f°[\text{reactants}])$
and we take the answer to Review Exercise 13.24 as a guide to deciding on which
reaction might be spontaneous, that is have a chance for $\Delta G = \Delta H - T\Delta S$ to be negative.
The data have been taken from Table 5.2.

(a) $\Delta H° = \{\Delta H_f°[CaCO_3(s)]\} - \{\Delta H_f°[CO_2(g)] + \Delta H_f°[CaO(s)]\}$
$\qquad \Delta H° = \{1 \text{ mol} \times (-1207 \text{ kJ/mol})\}$
$\qquad\qquad - \{1 \text{ mol} \times (-394 \text{ kJ/mol}) + 1 \text{ mol} \times (-635.5 \text{ kJ/mol})\}$
$\qquad \Delta H° = -178 \text{ kJ}$ ∴ favored.

(b) $\Delta H° = \{\Delta H_f°[C_2H_6(g)]\} - \{\Delta H_f°[C_2H_2(g)] + 2\Delta H_f°[H_2(g)]\}$

$\Delta H° = \{1 \text{ mol} \times (-84.5 \text{ kJ/mol})\}$
$- \{1 \text{ mol} \times (227 \text{ kJ/mol}) + 2 \text{ mol} \times (0.0 \text{ kJ/mol})\}$

$\Delta H° = -311 \text{ kJ} \therefore$ favored.

(c) $\Delta H° = \{\Delta H_f°[Fe_2O_3(s)] + 3\Delta H_f°[Ca(s)]\}$
$- \{2\Delta H_f°[Fe(s)] + 3\Delta H_f°[CaO(s)]\}$

$\Delta H° = \{1 \text{ mol} \times (-822.2 \text{ kJ/mol}) + 3 \text{ mol} \times (0.0 \text{ kJ/mol})\}$
$- \{2 \text{ mol} \times (0.0 \text{ kJ/mol}) + 3 \text{ mol} \times (-635.5 \text{ kJ/mol})\}$

$\Delta H° = +1084.3 \text{ kJ} \therefore$ not favorable from the standpoint of enthalpy alone.

(d) $\Delta H° = \{\Delta H_f°[H_2O(\ell)] + \Delta H_f°[CaO(s)]\} - \{\Delta H_f°[Ca(OH)_2(s)]\}$

$\Delta H° = \{1 \text{ mol} \times (-286 \text{ kJ/mol}) + 1 \text{ mol} \times (-635.5 \text{ kJ/mol})\}$
$- \{1 \text{ mol} \times (-986.6 \text{ kJ/mol})\}$

$\Delta H° = +65.1 \text{ kJ} \therefore$ not favored from the standpoint of enthalpy alone.

(e) $\Delta H° = \{2\Delta H_f°[HCl(g)] + \Delta H_f°[Na_2SO_4(s)]\}$
$- \{2\Delta H_f°[NaCl(s)] + \Delta H_f°[H_2SO_4(\ell)]\}$

$\Delta H° = \{2 \text{ mol} \times (-92.5 \text{ kJ/mol}) + 1 \text{ mol} \times (-1384.49 \text{ kJ/mol})\}$
$- \{2 \text{ mol} \times (-413 \text{ kJ/mol}) + 1 \text{ mol} \times (-811.8 \text{ kJ/mol})\}$

$\Delta H° = +68 \text{ kJ} \therefore$ not favored from the standpoint of enthalpy alone.

13.29 (a)   negative   (b)   negative   (c)   positive
       (d)   negative   (e)   negative   (f)   positive

13.34 The probability is given by the number of possibilities that lead to the desired arrangement, divided by the total number of possible arrangements.

We list each of the possible results, heads (H) or tails (T) for each of the coins, and systematically write down all of the distinct arrangements:

There is only one arrangement that gives four heads: HHHH.
There is only one arrangement that gives four tails: TTTT
Four distinct arrangements can lead to three heads and one tail:
    HHHT, THHH, HTHH, HHTH
There are similarly four distinct arrangements that lead to three tails and one head:
    TTTH, HTTT, THTT, TTHT
There are six distinct arrangements that can lead to two heads and two tails:
    HHTT, TTHH, THHT, THTH, HTTH, HTHT

Hence, the probability of all heads (HHHH) is 1 in 16 or $1/16 = 0.0625$.
The probability of two heads and two tails is 6 in 16 or $6/16 = 0.375$.

13.36 (a) negative – since the number of moles of gaseous material decreases.
     (b) negative – since the number of moles of gaseous material decreases.
     (c) negative – since the number of moles of gas decreases.
     (d) positive – since a gas appears where there formerly was none.

13.41 $\Delta S° = (\text{sum } S°[\text{products}]) - (\text{sum } S°[\text{reactants}])$

(a) $\Delta S° = \{2S°[NH_3(g)]\} - \{3S°[H_2(g)] + S°[N_2(g)]\}$

$\Delta S° = \{2 \text{ mol} \times (192.5 \text{ J mol}^{-1} \text{ K}^{-1})\} - \{3 \text{ mol} \times (130.6 \text{ J mol}^{-1} \text{ K}^{-1})$
$+ 1 \text{ mol} \times (191.5 \text{ J mol}^{-1} \text{ K}^{-1})\}$

$\Delta S° = -198.3 \text{ J/K} \therefore$ not spontaneous from the standpoint of entropy.

(b) $\Delta S° = \{S°[CH_3OH(\ell)]\} - \{2S°[H_2(g)] + S°[CO(g)]\}$

$\Delta S° = \{1 \text{ mol} \times (126.8 \text{ J mol}^{-1} \text{ K}^{-1})\}$
$- \{2 \text{ mol} \times (130.6 \text{ J mol}^{-1} \text{ K}^{-1}) + 1 \text{ mol} \times (197.9 \text{ J mol}^{-1} \text{ K}^{-1})\}$

$\Delta S° = -332.3 \text{ J/K} \therefore$ not favored from the standpoint of entropy alone.

(c) $\Delta S° = \{6S°[H_2O(g)] + 4S°[CO_2(g)]\} - \{7S°[O_2(g)] + 2S°[C_2H_6(g)]\}$

$\Delta S° = \{6 \text{ mol} \times (188.7 \text{ J mol}^{-1} \text{ K}^{-1}) + 4 \text{ mol} \times (213.6 \text{ J mol}^{-1} \text{ K}^{-1})\}$
$- \{7 \text{ mol} \times (205.0 \text{ J mol}^{-1} \text{ K}^{-1}) + 2 \text{ mol} \times (229.5 \text{ J mol}^{-1} \text{ K}^{-1})\}$

$\Delta S° = +92.6 \text{ J/K} \therefore$ favorable from the standpoint of entropy alone.

(d) $\Delta S° = \{2S°[H_2O(\ell)] + S°[CaSO_4(s)]\}$
$- \{S°[H_2SO_4(\ell)] + S°[Ca(OH)_2(s)]\}$

$\Delta S° = \{2 \text{ mol} \times (69.96 \text{ J mol}^{-1} \text{ K}^{-1}) + 1 \text{ mol} \times (107 \text{ J mol}^{-1} \text{ K}^{-1})\}$
$- \{1 \text{ mol} \times (157 \text{ J mol}^{-1} \text{ K}^{-1}) + 1 \text{ mol} \times (76.1 \text{ J mol}^{-1} \text{ K}^{-1})\}$

$\Delta S° = +14 \text{ J/K} \therefore$ favorable from the standpoint of entropy alone.

(e) $\Delta S° = \{2S°[N_2(g)] + S°[SO_2(g)]\} - \{2S°[N_2O(g)] + S°[S(s)]\}$

$\Delta S° = \{2 \text{ mol} \times (191.5 \text{ J mol}^{-1} \text{ K}^{-1}) + 1 \text{ mol} \times (248 \text{ J mol}^{-1} \text{ K}^{-1})\}$
$- \{2 \text{ mol} \times (220.0 \text{ J mol}^{-1} \text{ K}^{-1}) + 1 \text{ mol} \times (31.8 \text{ J mol}^{-1} \text{ K}^{-1})\}$

$\Delta S° = +159 \text{ J/K} \therefore$ favorable from the standpoint of entropy alone.

13.43 The entropy change that is designated $\Delta S_f°$ is that which corresponds to the reaction in which one mole of a substance is formed from elements in their standard states. Since the value is understood to correspond to the reaction forming one mole of a single pure substance, the units may be written either $J\ K^{-1}$ or $J\ mol^{-1}\ K^{-1}$.

(a) $2C(s) + 2H_2(g) \rightarrow C_2H_4(g)$

$\Delta S° = \{S°[C_2H_4(g)]\} - \{2S°[C(s)] + 2S°[H_2(g)]\}$

$\Delta S° = \{1\ mol \times (219.8\ J\ mol^{-1}\ K^{-1})\}$
$\qquad - \{2\ mol \times (5.69\ J\ mol^{-1}\ K^{-1}) + 2\ mol \times (130.6\ J\ mol^{-1}\ K^{-1})\}$

$\Delta S° = -52.8\ J/K$ or $-52.8\ J\ mol^{-1}\ K^{-1}$

(b) $N_2(g) + 1/2O_2(g) \rightarrow N_2O(g)$

$\Delta S° = \{S°[N_2O(g)]\} - \{S°[N_2(g)] + 1/2S°[O_2(g)]\}$

$\Delta S° = \{1\ mol \times (220.0\ J\ mol^{-1}\ K^{-1})\} - \{1\ mol \times (191.5\ J\ mol^{-1}\ K^{-1})$
$\qquad + 1/2\ mol \times (205.0\ J\ mol^{-1}\ K^{-1})\}$

$\Delta S° = -74.0\ J/K$ or $-74.0\ J\ mol^{-1}\ K^{-1}$

(c) $Na(s) + 1/2Cl_2(g) \rightarrow NaCl(s)$

$\Delta S° = \{S°[NaCl(s)]\} - \{1/2S°[Cl_2(g)] + S°[Na(s)]\}$

$\Delta S° = \{1\ mol \times (72.38\ J\ mol^{-1}\ K^{-1})\} - \{1/2\ mol \times 223.0\ J\ mol^{-1}\ K^{-1})$
$\qquad + 1\ mol \times (51.0\ J\ mol^{-1}\ K^{-1})\}$

$\Delta S° = -90.1\ J/K$ or $-90.1\ J\ mol^{-1}\ K^{-1}$

(d) $Ca(s) + S(s) + 3O_2(g) + 2H_2(g) \rightarrow CaSO_4 \cdot 2H_2O(s)$

$\Delta S° = \{S°[CaSO_4 \cdot 2H_2O(s)]\} - \{2S°[H_2(g)] + 3S°[O_2(g)] + S°[S(s)]$
$\qquad\qquad + S°[Ca(s)]\}$

$\Delta S° = \{1\ mol \times (194.0\ J\ mol^{-1}\ K^{-1})\} - \{2\ mol \times (130.6\ J\ mol^{-1}\ K^{-1})$
$\qquad + 3\ mol \times (205.0\ J\ mol^{-1}\ K^{-1}) + 1\ mol \times (31.8\ J\ mol^{-1}\ K^{-1})$
$\qquad + 1\ mol \times (41.4\ J\ mol^{-1}\ K^{-1})\}$

$\Delta S° = -755.4\ J/K$ or $-755.4\ J\ mol^{-1}\ K^{-1}$

(e) $2H_2(g) + 2C(s) + O_2(g) \rightarrow HC_2H_3O_2(\ell)$

$\Delta S° = \{S°[HC_2H_3O_2(\ell)]\} - \{2S°[H_2(g)] + 2S°[C(s)] + S°[O_2(g)]\}$

$\Delta S° = \{1 \text{ mol} \times (160 \text{ J mol}^{-1}\text{ K}^{-1})\} - \{2 \text{ mol} \times (130.6 \text{ J mol}^{-1}\text{ K}^{-1})$
$\qquad + 2 \text{ mol} \times (5.69 \text{ J mol}^{-1}\text{ K}^{-1}) + 1 \text{ mol} \times (205.0 \text{ J mol}^{-1}\text{ K}^{-1})\}$

$\Delta S° = -318 \text{ J/K or} -318 \text{ J mol}^{-1}\text{ K}^{-1}$

13.45 $\Delta S° = (\text{sum } S°[\text{products}]) - (\text{sum } S°[\text{reactants}])$

$\Delta S° = \{2S°[HNO_3(\ell)] + S°[NO(g)]\} - \{3S°[NO_2(g)] + S°[H_2O(\ell)]\}$

$\Delta S° = \{2 \text{ mol} \times (155.6 \text{ J mol}^{-1}\text{ K}^{-1}) + 1 \text{ mol} \times (210.6 \text{ J mol}^{-1}\text{ K}^{-1})\}$
$\qquad - \{3 \text{ mol} \times (240.5 \text{ J mol}^{-1}\text{ K}^{-1}) + 1 \text{ mol} \times (69.96 \text{ J mol}^{-1}\text{ K}^{-1})\}$

$\Delta S° = -269.7 \text{ J/K}$

13.50   The quantity $\Delta G_f°$ applies to the equation in which one mole of pure phosgene is produced from the naturally occurring forms of the elements:

$$C(s) + 1/2O_2(g) + Cl_2(g) \rightarrow COCl_2(g), \ \Delta G_f° = ?$$

We can determine $\Delta G_f°$ if we can find values for $\Delta H_f°$ and $\Delta S_f°$, because:

$\Delta G° = \Delta H° - T\Delta S°$

The value of $\Delta S_f°$ is determined using $S°$ for phosgene in the following way:

$\Delta S_f° = \{S°[COCl_2(g)]\} - \{S°[C(s)] + 1/2S°[O_2(g)] + S°[Cl_2(g)]\}$

$\Delta S_f° = \{1 \text{ mol} \times (284 \text{ J mol}^{-1}\text{ K}^{-1})\} - \{1 \text{ mol} \times (5.69 \text{ J mol}^{-1}\text{ K}^{-1})$
$\qquad + 1/2 \text{ mol} \times (205.0 \text{ J mol}^{-1}\text{ K}^{-1}) + 1 \text{ mol} \times (223.0 \text{ J mol}^{-1}\text{ K}^{-1})\}$

$\Delta S_f° = -47 \text{ J mol}^{-1}\text{ K}^{-1} \text{ or} -47 \text{ J/K}$

$\Delta G_f° = \Delta H_f° - T\Delta S_f° = -223 \text{ kJ/mol} - (298 \text{ K})(-0.047 \text{ kJ/mol K})$
$\qquad = -209 \text{ kJ/mol}$

13.52 $\Delta G° = (\text{sum } \Delta G_f°[\text{products}]) - (\text{sum } \Delta G_f°[\text{reactants}])$

(a) $\Delta G° = \{\Delta G_f°[H_2SO_4(\ell)]\} - \{\Delta G_f°[H_2O(\ell)] + \Delta G_f°[SO_3(g)]\}$

$\Delta G° = \{1 \text{ mol} \times (-689.9 \text{ kJ/mol})\} - \{1 \text{ mol} \times (-237.2 \text{ kJ/mol})$
$\qquad + 1 \text{ mol} \times (-370 \text{ kJ/mol})\}$

$\Delta G° = -83 \text{ kJ}$

(b) $\Delta G° = \{2\Delta G_f°[NH_3(g)] + \Delta G_f°[H_2O(\ell)] + \Delta G_f°[CaCl_2(s)]\}$
$- \{\Delta G_f°[CaO(s)] + 2\Delta G_f°[NH_4Cl(s)]\}$

$\Delta G° = \{2\text{ mol} \times (-16.7\text{ kJ/mol}) + 1\text{ mol} \times (-237.2\text{ kJ/mol})$
$+ 1\text{ mol} \times (-750.2\text{ kJ/mol})\} - \{1\text{ mol} \times (-604.2\text{ kJ/mol})$
$+ 2\text{ mol} \times (-203.9\text{ kJ/mol})\}$

$\Delta G° = -8.8\text{ kJ}$

(c) $\Delta G° = \{\Delta G_f°[H_2SO_4(\ell)] + \Delta G_f°[CaCl_2(s)]\} - \{\Delta G_f°[CaSO_4(s)]$
$+ 2\Delta G_f°[HCl(g)]\}$

$\Delta G° = \{1\text{ mol} \times (-689.9\text{ kJ/mol}) + 1\text{ mol} \times (-750.2\text{ kJ/mol})\}$
$- \{1\text{ mol} \times (-1320.3\text{ kJ/mol}) + 2\text{ mol} \times (-95.27\text{ kJ/mol})\}$

$\Delta G° = +70.7\text{ kJ}$

(d) $\Delta G° = \{\Delta G_f°[C_2H_5OH(\ell)]\} - \{\Delta G_f°[H_2O(g)] + \Delta G_f°[C_2H_4(g)]\}$

$\Delta G° = \{1\text{ mol} \times (-174.8\text{ kJ/mol})\} - \{1\text{ mol} \times (-228.6\text{ kJ/mol})$
$+ 1\text{ mol} \times (68.12\text{ kJ/mol})\}$

$\Delta G° = -14.3\text{ kJ}$

(e) $\Delta G° = \{2\Delta G_f°[H_2O(\ell)] + \Delta G_f°[SO_2(g)] + \Delta G_f°[CaSO_4(s)]\}$
$- \{2\Delta G_f°[H_2SO_4(\ell)] + \Delta G_f°[Ca(s)]\}$

$\Delta G° = \{2\text{ mol} \times (-237.2\text{ kJ/mol}) + 1\text{ mol} \times (-300\text{ kJ/mol})$
$+ 1\text{ mol} \times (-1320.3\text{ kJ/mol})\} - \{2\text{ mol} \times (-689.9\text{ kJ/mol})$
$+ 1\text{ mol} \times (0.0\text{ kJ/mol})\}$

$\Delta G° = -715\text{ kJ}$

13.54 $CaSO_4 \cdot 1/2 H_2O(s) + 3/2 H_2O(\ell) \rightarrow CaSO_4 \cdot 2H_2O(s)$

$\Delta G° = (\text{sum } \Delta G_f°[\text{products}]) - (\text{sum } \Delta G_f°[\text{reactants}])$

$\Delta G° = \{\Delta G_f°[CaSO_4 \cdot 2H_2O(s)]\} -$
$\{\Delta G_f°[CaSO_4 \cdot 1/2 H_2O(s)] + 3/2\Delta G_f°[H_2O(\ell)]\}$

$\Delta G° = \{1\text{ mol} \times (-1795.7\text{ kJ/mol})\}$
$- \{1\text{ mol} \times (-1435.2\text{ kJ/mol}) + 1.5\text{ mol} \times (-237.2\text{ kJ/mol})\}$

$\Delta G° = -4.7\text{ kJ}$

13.56 Multiply the reverse of the second equation by 2 (remembering to multiply the associated free energy change by –2), and add the result to the first equation:

$4NO(g) \rightarrow 2N_2O(g) + O_2(g)$, $\quad \Delta G° = -139.56$ kJ
$4NO_2(g) \rightarrow 4NO(g) + 2O_2(g)$, $\quad \Delta G° = +139.40$ kJ

$4NO_2 \rightarrow 3O_2(g) + 2N_2O(g)$, $\quad \Delta G° = -0.16$ kJ

This result is the reverse of the desired reaction, which must then have $\Delta G° = +0.16$ kJ

13.58 glucose + phosphate $\rightarrow$ glucose–phosphate + H_2O $\quad \Delta G = 13.13$ kJ
ATP + H_2O $\rightarrow$ ADP + phosphate $\quad \Delta G = -32.22$ kJ

Add these two reactions together to get:

glucose + ATP $\rightarrow$ glucose–phosphate + ADP $\quad \Delta G = -19.09$ kJ

13.63 The maximum work obtainable from a reaction is equal in magnitude to the value of ΔG for the reaction. Thus, we need only determine $\Delta G°$ for the process:

$\Delta G° = $ (sum $\Delta G_f°$[products]) – (sum $\Delta G_f°$[reactants])
$\Delta G° = \{3\Delta G_f°[H_2O(g)] + 2\Delta G_f°[CO_2(g)]\} - \{3\Delta G_f°[O_2(g)] + \Delta G_f°[C_2H_5OH(\ell)]\}$
$\Delta G° = \{3 \text{ mol} \times (-228.6 \text{ kJ/mol}) + 2 \text{ mol} \times (-394.4 \text{ kJ/mol})\}$
$\qquad - \{3 \text{ mol} \times (0.0 \text{ kJ/mol}) \; 1 \text{ mol} \times (-174.8 \text{ kJ/mol})\}$
$\Delta G° = -1299.8$ kJ

13.65 From Example 13.5 we see that the combustion of one mole of ethanol produces –1299.8 kJ of energy. In Example 13.6 we determined that the combustion of one mole of octane produces –5307 kJ. We need to calculate the amount of energy produced when one gallon of each of these fuels is burned;

$\text{\# mol ethanol} = \left(3.78 \times 10^3 \text{ mL ethanol}\right)\left(\frac{0.7893 \text{ g ethanol}}{1 \text{ mL ethanol}}\right)\left(\frac{1 \text{ mole ethanol}}{46.07 \text{ g ethanol}}\right)$

$\qquad = 64.8 \text{ moles ethanol}$

$\text{\# kJ} = (64.8 \text{ moles ethanol})\left(\frac{-1299.8 \text{ kJ}}{1 \text{ mole ethanol}}\right) = 8.42 \times 10^4 \text{ kJ}$

$$\text{\# mol octane} = \left(3.78 \times 10^3 \text{ mL octane}\right)\left(\frac{0.7025 \text{ g octane}}{1 \text{ mL octane}}\right)\left(\frac{1 \text{ mole octane}}{114.23 \text{ g octane}}\right)$$

$$= 23.2 \text{ moles octane}$$

$$\text{\# kJ} = (23.2 \text{ moles octane})\left(\frac{-5307 \text{ kJ}}{1 \text{ mole octane}}\right) = 1.23 \times 10^5 \text{ kJ}$$

In spite of the large number of moles of ethanol in one gallon of liquid, the energy produced from the combustion of a gallon of octane is greater than the amount produced when one gallon of ethanol is burned.

13.71 At equilibrium, $\Delta G = 0 = \Delta H - T\Delta S$

$T_{eq} = \Delta H/\Delta S$, and assuming that ΔS is independent of temperature, we have:

$$T_{eq} = (31.4 \times 10^3 \text{ J mol}^{-1}) \div (94.2 \text{ J mol}^{-1} \text{ K}^{-1}) = 333 \text{ K}$$

13.73 At equilibrium, $\Delta G = 0 = \Delta H - T\Delta S$

Thus $\Delta H = T\Delta S$, and if we assume that both ΔH and ΔS are independent of temperature, we have:

$$\Delta S = \Delta H/T_{eq} = (37.7 \times 10^3 \text{ J/mol}) \div (99.3 + 273.15 \text{ K})$$

$$\Delta S = 101.2 \text{ J mol}^{-1} \text{ K}^{-1}$$

13.75 The reaction is spontaneous if its associated value for $\Delta G°$ is negative.

$\Delta G° = (\text{sum } \Delta G_f°[\text{products}]) - (\text{sum } \Delta G_f°[\text{reactants}])$

$\Delta G° = \{\Delta G_f°[HC_2H_3O_2(\ell)] + \Delta G_f°[H_2O(\ell)] + \Delta G_f°[NO(g)] + \Delta G_f°[NO_2(g)]\}$
$\qquad\qquad - \{\Delta G_f°[C_2H_4(g)] + 2\Delta G_f°[HNO_3(\ell)]\}$

$\Delta G° = \{1 \text{ mol} \times (-392.5 \text{ kJ/mol}) + 1 \text{ mol} \times (-237.2 \text{ kJ/mol})$
$\qquad\qquad + 1 \text{ mol} \times (86.69 \text{ kJ/mol}) + 1 \text{ mol} \times (51.84 \text{ kJ/mol})\}$
$\qquad\qquad - \{1 \text{ mol} \times (68.12 \text{ kJ/mol}) + 2 \text{ mol} \times (-79.9 \text{ kJ/mol})\}$

$\Delta G° = -399.5 \text{ kJ}$

Yes, the reaction is spontaneous.

13.80 As the temperature is raised, $\Delta G'$ will become less negative, if $\Delta H°$ is negative and $\Delta S°$ is negative. Accordingly, less product will be present at equilibrium.

13.84 We can calculate the work of the expanding gas ($P\Delta V$) if we can calculate the change in volume ΔV. Since the initial volume is given (5.00 L), we need only to calculate the final volume. For this, it is first necessary to determine the value of n, the number of moles of gas.

$$n = \frac{PV}{RT} = \frac{(4.00 \text{ atm})(5.00 \text{ L})}{\left(0.0821 \frac{\text{L atm}}{\text{mol K}}\right)(298 \text{ K})} = 0.817 \text{ mol}$$

$$V_2 = \frac{nRT}{P_2} = \frac{(0.817 \text{ mol})\left(0.0821 \frac{\text{L atm}}{\text{mol K}}\right)(298 \text{ K})}{1 \text{ atm}} = 20.0 \text{ L}$$

$$\Delta V = (20.0 \text{ L} - 5.00 \text{ L}) = 15.0 \text{ L}$$

The work of gas expansion against a constant pressure of 1 atm is then given by the quantity $-P\Delta V$: $w = -(1 \text{ atm})(15.0 \text{ L}) = -15.0 \text{ L atm}$

13.85 As in question 13.84, we solve for the volume after the first expansion to a pressure of 2 atm:

$$V_2 = \frac{nRT}{P_2} = \frac{(0.817 \text{ mol})\left(0.0821 \frac{\text{L atm}}{\text{mol K}}\right)(298 \text{ K})}{2 \text{ atm}} = 10.0 \text{ L}$$

The work after the first stage of expansion is therefore:
$w_1 = -(2 \text{ atm})(10.0 \text{ L} - 5.00 \text{ L}) = -10.0 \text{ L atm}.$
The work performed during the second stage of expansion is:
$w_2 = -(1 \text{ atm})(20.0 \text{ L} - 10.0 \text{ L}) = -10.0 \text{ L atm}.$
The sum for the whole stepwise process is $w_1 + w_2 = -20.0 \text{ L atm}$.

13.87 First, review the information provided. Since the salt dissolved, the process is spontaneous, and the sign for ΔG is negative. The temperature went down indicating that this is an endothermic process and ΔH must be positive.. Dissolving the solid salt increases the disorder of the system which indicates ΔS is positive. The general equation from which we must work is $\Delta G = \Delta H - T\Delta S$. Using this equation, the magnitude of $T\Delta S$ must be larger than the magnitude of ΔH in order to obtain a negative value for ΔG.

13.92 This requires the breaking of three N–H single bonds:

$$NH_3 \rightarrow N + 3H$$

The enthalpy of atomization of NH_3 is thus three times the average N–H single bond energy: $3 \times 391 \text{ kJ/mol} = 1.17 \times 10^3 \text{ kJ/mol}$

13.94 $\Delta H_f°[C_2H_4(g)]$ refers to the enthalpy change under standard conditions for the following reaction:

$$2C_{graphite} + 2H_2(g) \rightarrow C_2H_4(g), \qquad \Delta H_f°[C_2H_4(g)] = 52.284 \text{ kJ/mol}$$

We can arrive at this net reaction in an equivalent way, namely, by vaporizing all of the necessary elements to give gaseous atoms, and then allowing the gaseous atoms to form all of the appropriate bonds. The overall enthalpy of formation by this route is numerically equal to that for the above reaction, and, conveniently, the enthalpy changes for each step are available in either Table 13.1 or Table 13.2:

$\Delta H_f° = \text{sum}(\Delta H_f°[\text{gaseous atoms}]) - \text{sum}(\text{average bond energies in the molecule})$
$\Delta H_f°[C_2H_4(g)] = 52.284 \text{ kJ/mol} = [2 \times 715.0 + 4 \times 218.0] - [4 \times 413 + C=C]$ from which we can calculate the C=C bond energy: 598 kJ/mol.

13.96 There are two C=S double bonds to be considered:
$\Delta H_f° = \text{sum}(\Delta H_f°[\text{gaseous atoms}]) - \text{sum}(\text{average bond energies in the molecule})$
$\Delta H_f°[CS_2(g)] = 115.3 \text{ kJ/mol} = [715.0 + 2 \times 274.7] - [2 \times C=S]$
The C=S double bond energy is therefore given by the equation:
$C=S = -(115.3 - 715.0 - 2 \times 274.7) \div 2 = 574.6 \text{ kJ/mol}$

13.98 $\Delta H_f° = \text{sum}(\Delta H_f°[\text{gaseous atoms}]) - \text{sum}(\text{average bond energies in the molecule})$
$\Delta H_f°[H_2S(g)] = -20.15 \text{ kJ/mol} = [274.7 + 2 \times 218.0] - [2 \times H-S]$
$H-S = (20.15 + 274.7 + 2 \times 218.0) \div 2 = 365.4 \text{ kJ/mol}$

13.100 We must consider two S=O bonds and two S–F bonds:
$\Delta H_f° = \text{sum}(\Delta H_f°[\text{gaseous atoms}]) - \text{sum}(\text{average bond energies in the molecule})$
$\Delta H_f°[SO_2F_2(g)] = -858 \text{ kJ/mol} = [274.7 + 2 \times 249.2 + 2 \times 78.91] -$
$$[2 \times 307.4 + 2 \times S=O]$$
$S=O = 587 \text{ kJ/mol}$

13.102 $\Delta H_f° = \text{sum}(\Delta H_f°[\text{gaseous atoms}]) - \text{sum}(\text{average bond energies in the molecule})$
$\Delta H_f°[CCl_4(g)] = [715.0 + 4 \times 121.0] - [4 \times 328] = -113 \text{ kJ/mol}$

CHAPTER FOURTEEN
Review Exercises

14.15 In cool weather, the rates of metabolic reactions of cold–blooded insects decrease, because of the affect of temperature on rate.

14.17 The low temperature causes the rate of metabolism to be very low.

14.22 The two rates of disappearance of SO_2Cl_2 are given by the slopes of tangents to the curve at the designated times:

Rate at t = 200 min = 1.0×10^{-4} moles L^{-1} min^{-1}

Rate at t = 600 min = 5.9×10^{-5} moles L^{-1} min^{-1}

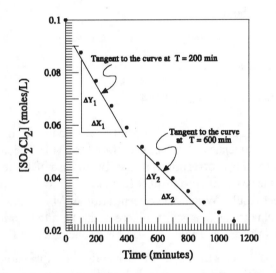

14.24 From the coefficients in the balanced equation we see that, for every mole of B that reacts, 2 mol of A are consumed, and three mol of C are produced. This means that A will be consumed twice as fast as B, and C will be produced three times faster than B is consumed.

rate of disappearance of A = 2(–0.30) = –0.60 mol L^{-1} s^{-1}

rate of appearance of C = 3(0.30) = 0.90 mol L^{-1} s^{-1}

14.26 (a) The rate of change in concentration is -2.5×10^{-6} mol L^{-1} s^{-1}, because the material disappears during the course of the reaction.

117

(b)   We rewrite the balanced chemical equation to make the problem easier to answer: $N_2O_5 \rightarrow 2NO_2 + 1/2O_2$. Thus, the rates of formation of NO_2 and O_2 will be, respectively, twice and one half the rate of disappearance of N_2O_5.

rate of formation of $NO_2 = 2(2.5 \times 10^{-6}) = 5.0 \times 10^{-6}$ mol L^{-1} s^{-1}

rate of formation of $O_2 = 1/2(2.5 \times 10^{-6}) = 1.3 \times 10^{-6}$ mol L^{-1} s^{-1}

14.35   rate $= (7.1 \times 10^9$ L^2 mol^{-2} $s^{-1})(1.0 \times 10^{-3}$ mol/L$)^2(3.4 \times 10^{-2}$ mol/L$)$
        rate $= 2.4 \times 10^2$ mol L^{-1} s^{-1}

14.37   In each case, the order with respect to a reactant is the exponent to which that reactant's concentration is raised in the rate law.

(a)   For $HCrO_4^-$, the order is 1.

       For HSO_3^-, the order is 2.

       For H^+, the order is 1.

(b)   The overall order is $1 + 2 + 1 = 4$.

14.40   On comparing the data of the first and second experiments, we find that, whereas the concentration of N is unchanged, the concentration of M has been doubled, causing a doubling of the rate. This corresponds to the fourth case in Table 14.3, and we conclude that the order of the reaction with respect to M is 1. In the second and third experiments, we have a different result. When the concentration of M is held constant, the concentration of N is tripled, causing an increase in the rate by a factor of nine. This constitutes the eighth case in Table 14.3, and we conclude that the order of the reaction with respect to N is 2. This means that the overall rate expression is: rate $= k[M][N]^2$ and we can solve for the value of k by substituting the appropriate data:

5.0×10^{-3} mol L^{-1} $s^{-1} = k \times [0.020$ mol/L$][0.010$ mol/L$]^2$
$k = 2.5 \times 10^3$ L^2 mol^{-2} s^{-1}

14.42 The reaction is first order in OCl^-, because an increase in concentration by a factor of two, while holding the concentration of I^- constant (compare the first and second experiments of the table), has caused an increase in rate by a factor of $2^1 = 2$. The order of reaction with respect to I^- is also 1, as is demonstrated by a comparison of the second and third experiments.

rate $= k[OCl^-][I^-]$
Using the last data set:
3.5×10^4 mol L^{-1} $s^{-1} = k[1.7 \times 10^{-3}$ mol/L$][3.4 \times 10^{-3}$ mol/L$]$
$k = 6.1 \times 10^9$ L mol^{-1} s^{-1}

14.44 Compare the first and second experiments. On doubling the ICl concentration, the rate is found to increase by a factor of $2 = 2^1$, and the order of the reaction with respect to ICl is 1 (case number four in Table 14.3). In the first and third experiments, the concentration of ICl is constant, whereas the concentration of H_2 in the first experiment is twice that in the third. This causes a change in the rate by a factor of 2 also, and the rate law is found to be: rate $= k[ICl][H_2]$. Using the data of the first experiment:
1.5×10^{-3} mol L^{-1} s$^{-1} = k[0.10$ mol $L^{-1}][0.10$ mol $L^{-1}]$
$k = 1.5 \times 10^{-1}$ L mol^{-1} s^{-1}

14.47 A graph of $\ln [SO_2Cl_2]_t$ versus t will yield a straight line if the data obeys a first-order rate law.

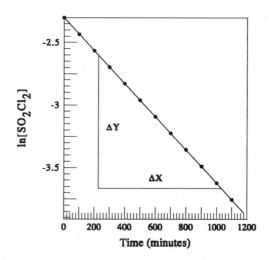

These data do yield a straight line when ln $[SO_2Cl_2]_t$ is plotted against the time, t. The slope of this line equals $-k$. Plotting the data provided and using linear regression to fit the data to a straight line yields a value of 1.32×10^{-3} min^{-1} for k.

14.49 (a) The time involved must be converted to a value in seconds:

1 hr $\times$ 3600 s/hr $= 3.6 \times 10^3$ s, and then we make use of equation 14.5, where x is taken to represent the desired SO_2Cl_2 concentration:

$$\ln \frac{0.0040 \text{ M}}{x} = (2.2 \times 10^{-5} \text{ s}^{-1})(3.6 \times 10^3 \text{ s})$$

$$x = 3.7 \times 10^{-3} \text{ M}$$

(b) The time is converted to a value having the units seconds

24 hr $\times$ 3600 s/hr $= 8.64 \times 10^4$ s, and then we use equation 14.5, where x is taken to represent the desired SO_2Cl_2 concentration:

$$\ln \frac{0.0040 \text{ M}}{x} = (2.2 \times 10^{-5} \text{ s}^{-1})(8.64 \times 10^4 \text{ s})$$

$$x = 6.0 \times 10^{-4} \text{ M}$$

14.51 Use equation 14.6, where the time is: 2.5×10^3 min $\times$ 60 s/min $= 1.5 \times 10^5$ s

$$\frac{1}{[B]_t} - \frac{1}{[B]_0} = kt$$

$$\frac{1}{\left[4.5 \times 10^{-4} \text{ M}\right]} - \frac{1}{[HI]_0} = \left(1.6 \times 10^{-3} \text{ L mol}^{-1} \text{ s}^{-1}\right)\left(1.5 \times 10^5 \text{ s}\right)$$

$$\frac{1}{[HI]_0} = 2.0 \times 10^3 \text{ L mol}^{-1}$$

$$[HI]_0 = 5.0 \times 10^{-4} \text{ M}$$

14.53 Use equation 14.5, taking time in minutes; 3 hr $= 180$ min.

$$\ln \frac{x}{5.0 \text{ mg}/_{kg}} = \left(4.26 \times 10^{-3} \text{ min}^{-1}\right)(180 \text{ min})$$

$x = 11$ mg/kg

14.59 $t = 0.693/k = 0.693/1.6 \times 10^{-3} \text{ s}^{-1} = 4.3 \times 10^2$ seconds

14.61 It requires approximately 500 min (as determined from the graph) for the concentration of SO_2Cl_2 to decrease from 0.100 M to 0.050 M, i.e., to decrease to half its initial concentration. Likewise, in another 500 minutes, the concentration decreases by half again, i.e. from 0.050 M to 0.025 M. This means that the half life of the reaction is independent of the initial concentration, and we conclude that the reaction is first order in SO_2Cl_2.

14.63 # half lives $= (2.0 \text{ hrs})\left(\dfrac{60 \text{ min}}{1 \text{ hr}}\right)\left(\dfrac{1 \text{ half life}}{15 \text{ min}}\right) = 8.0$ half lives

Eight half lives correspond to the following fraction of original material remaining:

Number of half lives	Fraction remaining
1	1/2
2	1/4
3	1/8
4	1/16
5	1/32
6	1/64
7	1/128
8	1/256

14.76 The graph is prepared exactly as in Example 14.11 of the text. The slope is found using linear regression, to be: -9.5×10^3 K. Thus -9.5×10^3 K $= -E_a/R$

$E_a = -(-9.5 \times 10^3 \text{ K})(8.314 \text{ J K}^{-1} \text{ mol}^{-1}) = 7.9 \times 10^4$ J/mol = 79 kJ/mol

Using equation 14.11, we proceed as follows:

$$\ln \frac{k_2}{k_1} = \frac{-E_a}{R}\left[\frac{1}{T_2} - \frac{1}{T_1}\right]$$

$$\ln\left[\frac{1.94 \times 10^{-3} \text{ L mol}^{-1} \text{ s}^{-1}}{2.88 \times 10^{-4} \text{ L mol}^{-1} \text{ s}^{-1}}\right] = \frac{-E_a}{8.314 \text{ J mol}^{-1} \text{ K}^{-1}}\left[\frac{1}{673 \text{ K}} - \frac{1}{593 \text{ K}}\right]$$

$$1.907 = \frac{2.00 \times 10^{-4} \text{ K}^{-1}}{8.314 \text{ J mol}^{-1} \text{ K}^{-1}} \times E_a$$

$E_a = 7.93 \times 10^4$ J/mol = 79.3 kJ/mol

14.78 Using equation 14.11 we have:

$$\ln \frac{k_2}{k_1} = \frac{-E_a}{R}\left[\frac{1}{T_2} - \frac{1}{T_1}\right]$$

$$\ln\left[\frac{1.0 \times 10^{-3}\ \text{L mol}^{-1}\ \text{s}^{-1}}{9.3 \times 10^{-5}\ \text{L mol}^{-1}\ \text{s}^{-1}}\right] = \frac{-E_a}{8.314\ \text{J mol}^{-1}\ \text{K}^{-1}}\left[\frac{1}{403\ \text{K}} - \frac{1}{373\ \text{K}}\right]$$

$$2.4 = \frac{2.00 \times 10^{-4}\ \text{K}^{-1}}{8.314\ \text{J mol}^{-1}\ \text{K}^{-1}} \times E_a$$

$E_a = 1.0 \times 10^5$ J/mol = 100 kJ/mol

Equation 14.9 states $k = A \exp\left(\dfrac{-E_a}{RT}\right)$

$$A = \frac{k}{\exp\left(\dfrac{-E_a}{RT}\right)}$$

$$= \frac{9.3 \times 10^{-5}\ \text{L mol}^{-1}\ \text{s}^{-1}}{\exp\left(\dfrac{-1.0 \times 10^5\ \text{J}/\text{mol}}{\left(8.314\ \text{J}/\text{mol K}\right)\left(373\ \text{K}\right)}\right)}$$

$$= 9.4 \times 10^9\ \text{L mol}^{-1}\ \text{s}^{-1}$$

14.80 Substituting into equation 14.11:

$$\ln \frac{k_2}{k_1} = \frac{-E_a}{R}\left[\frac{1}{T_2} - \frac{1}{T_1}\right]$$

$$\ln\left[\frac{3.75 \times 10^{-2}\ \text{s}^{-1}}{2.1 \times 10^{-3}\ \text{s}^{-1}}\right] = \frac{-E_a}{8.314\ \text{J mol}^{-1}\ \text{K}^{-1}}\left[\frac{1}{298\ \text{K}} - \frac{1}{273\ \text{K}}\right]$$

$(3.70 \times 10^{-5}\ \text{mol/J})(E_a) = 2.88$

$E_a = 7.8 \times 10^4$ J/mol = 78 kJ/mol

14.82

$$\ln \frac{k_2}{k_1} = \frac{-E_a}{R} \left[\frac{1}{T_2} - \frac{1}{T_1} \right]$$

$$\ln \left[\frac{2}{1} \right] = \frac{-E_a}{8.314 \text{ J mol}^{-1} \text{ K}^{-1}} \left[\frac{1}{308 \text{ K}} - \frac{1}{298 \text{ K}} \right]$$

$(1.31 \text{ X } 10^{-5} \text{ mol/J})(E_a) = 0.693$

$E_a = 5.29 \text{ X } 10^4 \text{ J/mol} = 52.9 \text{ kJ/mol}$

14.84 Substitute into equation 14.11:

$$\ln \frac{k_2}{k_1} = \frac{-E_a}{R} \left[\frac{1}{T_2} - \frac{1}{T_1} \right]$$

$$\ln \left[\frac{k_2}{6.2 \text{ X } 10^{-5} \text{ s}^{-1}} \right] = \frac{-108 \text{ X } 10^3 \text{ J mol}^{-1}}{8.314 \text{ J mol}^{-1} \text{ K}^{-1}} \left[\frac{1}{318 \text{ K}} - \frac{1}{308 \text{ K}} \right]$$

$\ln (k_2/6.2 \text{ X } 10^{-5} \text{ s}^{-1}) = 1.33$

$k_2 = 6.2 \text{ X } 10^{-5} \text{ s}^{-1} \times \text{antiln}(1.33) = 2.3 \text{ X } 10^{-4} \text{ s}^{-1}$

14.88 Adding all of the steps gives:

$2NO + 2H_2 \rightarrow N_2 + 2H_2O$

14.94 The predicted rate law is based on the rate–determining step:

$\text{rate} = k[NO_2]^2$

14.102 $k = 0.693/12.5 \text{ y} = 5.54 \text{ X } 10^{-2} \text{ y}^{-1}$

$$\ln \frac{1}{0.1} = \left(5.54 \text{ X } 10^{-2} \text{ y}^{-1} \right) \times t$$

$t = 41.6 \text{ y}$

14.104 (a)   The order with respect to [B] is:

$$\left(\frac{0.040 \text{ mol L}^{-1}}{0.030 \text{ mol L}^{-1}}\right)^{n} = \left(\frac{-0.0334 \text{ mol L}^{-1} \text{ s}^{-1}}{-0.0188 \text{ mol L}^{-1} \text{ s}^{-1}}\right) = 1.78 \qquad \therefore \text{ n} = 2$$

The order with respect to [A] is:

$$\left(\frac{0.025 \text{ mol L}^{-1}}{0.020 \text{ mol L}^{-1}}\right)^{n} = \left(\frac{-0.0188 \text{ mol L}^{-1} \text{ s}^{-1}}{-0.0150 \text{ mol L}^{-1} \text{ s}^{-1}}\right) = 1.25 \qquad \therefore \text{ n} = 1$$

$$\text{rate} = k[A]^{1}[B]^{2}$$

(b)   One fact that is not immediately obvious, but that you may have noticed throughout this chapter, is that the value of a rate constant is always positive. Hence, we will take the absolute value of the initial rate for this calculation. Using the first rate value, we substitute into the rate expression from part (a).

$$0.0150 \text{ mol L}^{-1} \text{ s}^{-1} = k(0.020 \text{ mol L}^{-1})^{1}(0.030 \text{ mol L}^{-1})^{2}$$

$$k = 8.33 \times 10^{2} \text{ L}^{2} \text{ mol}^{-2} \text{ s}^{-1}$$

15.6 (a) $K_c = \dfrac{[POCl_3]^2}{[PCl_3]^2[O_2]}$

(d) $K_c = \dfrac{[NO_2]^2[H_2O]^8}{[N_2H_4][H_2O_2]^6}$

(b) $K_c = \dfrac{[SO_2]^2[O_2]}{[SO_3]^2}$

(e) $K_c = \dfrac{[SO_2][HCl]^2}{[SOCl_2][H_2O]}$

(c) $K_c = \dfrac{[NO]^2[H_2O]^2}{[N_2H_4][O_2]^2}$

15.8 (a) $K_c = \dfrac{\left[Ag(NH_3)_2{}^+\right]}{\left[Ag^+\right][NH_3]^2}$

(b) $K_c = \dfrac{\left[Cd(SCN)_4{}^{2-}\right]}{\left[Cd^{2+}\right]\left[SCN^-\right]^4}$

(c) $K_c = \dfrac{\left[H_3O^+\right]\left[ClO^-\right]}{[HClO]}$

15.10 (a) $K_c = \dfrac{[HCl]^2}{[H_2][Cl_2]}$ (b) $K_c = \dfrac{[HCl]}{[H_2]^{1/2}[Cl_2]^{1/2}}$

K_c for reaction (b) is the square root of K_c for reaction (a).

15.13 (a) $K_p = \dfrac{\left(P_{POCl_3}\right)^2}{\left(P_{PCl_3}\right)^2\left(P_{O_2}\right)}$

(c) $K_p = \dfrac{\left(P_{NO}\right)^2\left(P_{H_2O}\right)^2}{\left(P_{N_2H_4}\right)\left(P_{O_2}\right)^2}$

(b) $K_p = \dfrac{\left(P_{SO_2}\right)^2\left(P_{O_2}\right)}{\left(P_{SO_3}\right)^2}$

(d) $K_p = \dfrac{\left(P_{NO_2}\right)^2\left(P_{H_2O}\right)^8}{\left(P_{N_2H_4}\right)\left(P_{H_2O_2}\right)^6}$

(e) $\quad K_p = \dfrac{(P_{SO_2})(P_{HCl})^2}{(P_{SOCl_2})(P_{H_2O})}$

15.17 The increasing tendency to go to completion is: (a) < (c) < (b), based on the relative magnitudes of K_c.

15.23 (a) $\Delta G° = \{2 \times \Delta G_f°[POCl_3(g)]\} - \{2 \times \Delta G_f°[PCl_3(g)] + 2 \times \Delta G_f°[O_2(g)]\}$

$\Delta G° = \{2 \text{ mol} \times (-1019 \text{ kJ/mol})\}$
$\qquad\qquad - \{2 \text{ mol} \times (-267.8 \text{ kJ/mol}) + 1 \text{ mol} \times (0 \text{ kJ/mol})\}$

$\Delta G° = -1502 \text{ kJ} = -1.502 \times 10^6 \text{ J}$

$-1.502 \times 10^6 \text{ J} = -RT\ln K_p = -(8.314 \text{ J/K mol})(298 \text{ K}) \times \ln K_p$

$\ln K_p = 606$ $\therefore \log K_p = 263$ and $K_p = 10^{263}$.

(b) $\Delta G° = \{2 \times \Delta G_f°[SO_2(g)] + 1 \times \Delta G_f°[O_2(g)]\} - \{2 \times \Delta G_f°[SO_3(g)]\}$

$\Delta G° = \{2 \text{ mol} \times (-300 \text{ kJ/mol}) + 1 \text{ mol} \times (0 \text{ kJ/mol})\} - \{2 \text{ mol} \times (-370 \text{ kJ/mol})\}$

$\Delta G° = 140 \text{ kJ} = 1.40 \times 10^5 \text{ J}$

$1.40 \times 10^5 \text{ J} = -RT\ln K_p = -(8.314 \text{ J/K mol})(298 \text{ K}) \times \ln K_p$

$\ln K_p = -56.5$ and $K_p = 2.90 \times 10^{-25}$

(c) $\Delta G° = \{2 \times \Delta G_f°[NO(g)] + 2 \times \Delta G_f°[H_2O(g)]\}$
$\qquad\qquad - \{1 \times \Delta G_f°[N_2H_4(g)] + 2 \times \Delta G_f°[O_2(g)]\}$

$\Delta G° = \{2 \text{ mol} \times (86.69 \text{ kJ/mol}) + 2 \text{ mol} \times (-228.6 \text{ kJ/mol})\}$
$\qquad\qquad - \{1 \text{ mol} \times (159.3 \text{ kJ/mol}) + 2 \text{ mol} \times (0.0 \text{ kJ/mol})\}$

$\Delta G° = -443.1 \text{ kJ} = -4.431 \times 10^5 \text{ J}$

$-4.431 \times 10^5 \text{ J} = -RT\ln K_p = -(8.314 \text{ J/K mol})(298 \text{ K}) \times \ln K_p$

$\ln K_p = 178.9$ and $K_p = 4.96 \times 10^{77}$

(d) $\Delta G° = \{2 \times \Delta G_f°[NO_2(g)] + 8 \times \Delta G_f°[H_2O(g)]\}$
$\qquad\qquad - \{1 \times \Delta G_f°[N_2H_4(g)] + 6 \times \Delta G_f°[H_2O_2(g)]\}$

$\Delta G° = [2 \text{ mol} \times (51.84 \text{ kJ/mol}) + 8 \text{ mol} \times (-228.6 \text{ kJ/mol})]$
$\qquad\qquad - [1 \text{ mol} \times (159.3 \text{ kJ/mol}) + 6 \text{ mol} \times (-105.6 \text{ kJ/mol})]$

$\Delta G° = -1250.8 \text{ kJ} = -1.2508 \times 10^6 \text{ J}$

$-1.2508 \times 10^6 \text{ J} = -RT\ln K_p = -(8.314 \text{ J/K mol})(298 \text{ K}) \times \ln K_p$

$\ln K_p = 504.85$ and $\log K_p = 219.21$ $\therefore K_p = 10^{219}$

15.25 $\Delta G^\circ = -RT \ln K_p$

$-9.67 \times 10^3 \text{ J} = -(8.314 \text{ J/K mol})(1273 \text{ K}) \times \ln K_p$

$\ln K_p = 0.914 \quad \therefore \quad K_p = 2.49$

$$Q = \frac{[N_2O][O_2]}{[NO_2][NO]} = \frac{(0.015)(0.0350)}{(0.0200)(0.040)} = 0.66$$

Since the value of Q is less than the value of K, the system is not at equilibrium and must shift to the right to reach equilibrium.

15.27 $\Delta G^\circ = -RT \ln K_p$

$-50.79 \times 10^3 \text{ J} = -(8.314 \text{ J K}^{-1} \text{ mol}^{-1})(298 \text{ K}) \times \ln K_p$

$\ln K_p = 20.50$

Taking the antiln of both sides of this equation gives: $K_p = 8.000 \times 10^8$

This is a favorable reaction, since the equilibrium lies far to the side favoring products and is worth studying as a method for methane production.

15.31 $\Delta G^\circ_{500} = -RT \ln K_p$

$\Delta G^\circ_{500} = -(8.314 \text{ J K}^{-1} \text{ mol}^{-1})(500 \text{ K}) \ln (6.25 \times 10^{-3})$

$\Delta G^\circ_{500} = 2.11 \times 10^4 \text{ J} = 21.1 \text{ kJ}$

15.33 First, we determine ΔH° by using the normal approach:

$\Delta H^\circ = (\text{sum } \Delta H_f^\circ[\text{products}]) - (\text{sum } \Delta H_f^\circ[\text{reactants}])$

$\Delta H^\circ = \{\Delta H_f^\circ[N_2(g)] + 2 \times \Delta H_f^\circ[CO_2(g)]\} - \{2 \times \Delta H_f^\circ[NO(g)] + 2 \times \Delta H_f^\circ[CO(g)]\}$

$\Delta H^\circ = \{1 \text{ mol} \times (0.0 \text{ kJ/mol}) + 2 \text{ mol} \times (-394 \text{ kJ/mol})\}$
$\qquad - \{2 \text{ mol} \times (90.4 \text{ kJ/mol}) + 2 \text{ mol} \times (-110 \text{ kJ/mol})\}$

$\Delta H^\circ = -749 \text{ kJ}$ for the formation of one mole of N_2.

$\Delta S^\circ = \{S^\circ[N_2(g)] + 2 \times S^\circ[CO_2(g)]\} - \{2 \times S^\circ[NO(g)] + 2 \times S^\circ[CO(g)]\}$

$\Delta S^\circ = \{1 \text{ mol} \times (191.5 \text{ J K}^{-1} \text{ mol}^{-1}) + 2 \text{ mol} \times (213.6 \text{ J K}^{-1} \text{ mol}^{-1})\}$
$\qquad - \{2 \text{ mol} \times (210.6 \text{ J K}^{-1} \text{ mol}^{-1}) + 2 \text{ mol} \times (197.9 \text{ J K}^{-1} \text{ mol}^{-1})\}$

$\Delta S^\circ = -198.3 \text{ J/K}$ for the formation of one mole of N_2.

$\Delta G°_T = \Delta H° - T\Delta S°$

$\Delta G°_{773} = -749 \text{ kJ} - (773 \text{ K})(-0.1983 \text{ kJ/K}) = -596 \text{ kJ/mol}$

$\Delta G°_T = -RT \ln K_p$

$-596 \text{ kJ/mol} = -(8.314 \text{ J K}^{-1} \text{ mol}^{-1})(773 \text{ K}) \ln K_p$

$\ln K_p = 92.7$

Taking the antiln of both sides of this equation gives:

$K_p = 1.82 \times 10^{40}$

15.36 $M = P/RT$

$$M = \frac{(745 \text{ torr})\left(\dfrac{1 \text{ atm}}{760 \text{ torr}}\right)}{\left(0.0821 \frac{\text{L atm}}{\text{mol K}}\right)(318 \text{ K})} = 0.0375 \text{ M}$$

15.38 $K_p = K_c \times (RT)^{\Delta n_g}$

If the value of $(RT)^{\Delta n_g}$ is unity, then the left and right sides of this equation become equal to one another. This can happen only if there is no change in the total number of gaseous substances on going from reactants to products. This happens in reactions (a) and (d) only.

15.39 $K_p = K_c \times (RT)^{\Delta n_g}$

$6.3 \times 10^{-3} = K_c[(0.0821 \text{ L atm K}^{-1} \text{ mol}^{-1})(498 \text{ K})]^{-2} = 5.98 \times 10^{-4} \times K_c$

$K_c = 11$

15.41 $K_p = K_c \times (RT)^{\Delta n_g}$

$K_p = 4.2 \times 10^{-4}[(0.0821 \text{ L atm K}^{-1} \text{ mol}^{-1})(773 \text{ K})]^{1} = 2.7 \times 10^{-2}$

15.43 $K_p = K_c \times (RT)^{\Delta n_g}$

$K_p = (0.40)[(0.0821 \text{ L atm K}^{-1} \text{ mol}^{-1})(1046 \text{ K})]^{-2} = 5.4 \times 10^{-5}$

15.46 In each case we get approximately 55.5 M:

(a)

$$\# \text{ mol } H_2O = (18.0 \text{ mL } H_2O)\left(\frac{1 \text{ g}}{1 \text{ mL}}\right)\left(\frac{1 \text{ mol } H_2O}{18.02 \text{ g } H_2O}\right) = 0.999 \text{ mol } H_2O$$

$$M = \left(\frac{0.999 \text{ mol } H_2O}{18.0 \text{ mL } H_2O}\right)\left(\frac{1000 \text{ mL}}{1 \text{ L}}\right) = 55.5 \text{ M}$$

(b)

$$\# \text{ mol } H_2O = (100.0 \text{ mL } H_2O)\left(\frac{1 \text{ g}}{1 \text{ mL}}\right)\left(\frac{1 \text{ mol } H_2O}{18.02 \text{ g } H_2O}\right) = 5.549 \text{ mol } H_2O$$

$$M = \left(\frac{5.549 \text{ mol } H_2O}{100.0 \text{ mL } H_2O}\right)\left(\frac{1000 \text{ mL}}{1 \text{ L}}\right) = 55.49 \text{ M}$$

(c)

$$\# \text{ mol } H_2O = (1.00 \text{ L } H_2O)\left(\frac{1000 \text{ mL}}{1 \text{ L}}\right)\left(\frac{1 \text{ g}}{1 \text{ mL}}\right)\left(\frac{1 \text{ mol } H_2O}{18.02 \text{ g } H_2O}\right)$$

$$= 55.5 \text{ mol } H_2O$$

$$M = \left(\frac{55.5 \text{ mol } H_2O}{1.00 \text{ L } H_2O}\right) = 55.5 \text{ M}$$

15.49 (a) $K_c = \dfrac{[CO]^2}{[O_2]}$ \qquad (c) $K_c = \dfrac{[CH_4][CO_2]}{[H_2O]^2}$

(b) $K_c = [H_2O][SO_2]$ \qquad (d) $K_c = \dfrac{[H_2O][CO_2]}{[HF]^2}$

15.51

	[HCl]	[HI]	[Cl₂]
I	0.100	–	–
C	–2x	+2x	+x
E	0.100–2x	+2x	+x

Note: Since the $I_2(s)$ has a constant concentration, it may be neglected.

$$K_c = \frac{[HI]^2[Cl_2]}{[HCl]^2} = 1.0 \times 10^{-34}$$

$$K_c = \frac{(2x)^2(x)}{(0.100 - 2x)^2} = 1.0 \times 10^{-34}$$

Because the value of K_c is so small, we make the simplifying assumption that $(0.100 - 2x) \approx 0.100$, and the above equation becomes:

$$K_c = \frac{[HI]^2[Cl_2]}{[HCl]^2} = 1.0 \times 10^{-34}$$

$$K_c = \frac{(2x)^2(x)}{(0.100)^2} = 1.0 \times 10^{-34}$$

$4x^3 = 1.6 \times 10^{-36}$; $\therefore$ $x = 7.37 \times 10^{-13}$ M, and the above assumption is seen to have been valid.

$[HI] = 2x = 1.47 \times 10^{-12}$ M

$[Cl_2] = x = 7.37 \times 10^{-13}$ M

$[HCl] = (0.100 - 2x) \approx 0.100$ M

15.54 (a)   The system shifts to the right to consume some of the added methane.
     (b)   The system shifts to the left to consume some of the added hydrogen.
     (c)   The system shifts to the right to make some more carbon disulfide.
     (d)   The system shifts to the left to decrease the amount of gas.
     (e)   The system shifts to the right to absorb some of the added heat.

15.56 (a)   right   (b)   left   (c)   left
     (d)   right   (e)   no effect   (f)   left

15.58 (a)   increase   (b)   increase
     (c)   increase   (d)   decrease

15.59 $K_c = \dfrac{[CH_3OH]}{[CO][H_2]^2} = \dfrac{(0.00261)}{(0.105)(0.250)^2} = 0.398$

15.60   The mass action expression for this equilibrium is:

$$K_c = \frac{[PCl_5]}{[PCl_3][Cl_2]} = 0.18$$

and the value for the ion product for this system is:

$$Q = \frac{(0.00500)}{(0.0420)(0.0240)} = 4.96$$

(a)   This is not the value of the equilibrium constant, and we conclude that the system is not at equilibrium.

(b)   Since the value of the ion product for this system is larger than that of the equilibrium constant, the system must shift to the left to reach equilibrium.

15.62   (a)   The mass action expression is:

$$K_p = \frac{\left(P_{NO_2}\right)^2}{\left(P_{N_2O_4}\right)} = 0.140 \text{ atm}$$

Solving the above expression for the partial pressure of NO_2, we get:

$$P_{NO_2} = \sqrt{P_{N_2O_4} \times K_p} = \sqrt{(0.250 \text{ atm})(0.140 \text{ atm})} = 0.187 \text{ atm}$$

(b) $P_{total} = P_{NO_2} + P_{N_2O_4} = 0.187 + 0.250 = 0.437 \text{ atm}$

15.64 $K_c = \dfrac{[CH_3OH]}{[CO][H_2]^2} = \dfrac{[CH_3OH]}{(0.180)(0.220)^2} = 0.500$

$[CH_3OH] = 4.36 \times 10^{-3} \text{ M}.$

15.66 If we substitute the initial concentrations of 0.0400 M for each component of this mixture into the ion product expression that results from the balanced equation in the text, we see that the ion product equals one. Since $Q > K_c$, the reaction will proceed from right to left as written, i.e., $[Br_2]$ and $[Cl_2]$ will decrease and $[BrCl]$ will increase.

	[BrCl]	[Br₂]	[Cl₂]
I	0.0400	0.0400	0.0400
C	+2x	–x	–x
E	0.0400+2x	0.0400–x	0.0400–x

Substituting the above values for equilibrium concentrations into the mass action expression gives:

$$K_c = \frac{[Br_2][Cl_2]}{[BrCl]^2} = \frac{(0.0400 - x)(0.0400 - x)}{(0.0400 + 2x)^2} = 0.145$$

Take the square root of both sides of this equation to get; $\frac{(0.0400 - x)}{(0.0400 + 2x)} = 0.381$. This equation is easily solved giving x = 0.0141 M. The substances then have the following concentrations at equilibrium: $[Cl_2] = [Br_2] = 0.0400 - 0.0141 = 0.0259$ M, $[BrCl] = 0.0400 + 2(0.0141) = 0.0682$ M.

15.68 If we substitute the initial concentrations for each component of this mixture into the ion product expression that results from the balanced equation in the text, we see that the ion product equals 0.375. Since $Q > K_c$ for this situation, the reaction will proceed from right to left, i.e., $[Br_2]$ and $[Cl_2]$ will decrease and $[BrCl]$ will increase.

	[BrCl]	[Br₂]	[Cl₂]
I	0.0400	0.0300	0.0200
C	+2x	–x	–x
E	0.0400+2x	0.0300–x	0.0200–x

Substituting the above values for equilibrium concentrations into the mass action expression gives:

$$K_c = \frac{[Br_2][Cl_2]}{[BrCl]^2} = \frac{(0.0300 - x)(0.0200 - x)}{(0.0400 + 2x)^2} = 0.145$$

Rearranging we get the following quadratic expression: $0 = 0.42x^2 - 0.0732x + 0.000368$. Solving and using the negative root gives $x = 0.0052$ M. The individual concentrations are: $[Cl_2] = 0.0200 - 0.0052 = 0.0148$ M, $[Br_2] = 0.0300 - 0.0052 = 0.0248$ M, $[BrCl] = 0.0400 + 2(0.0052) = 0.0504$ M.

15.70

	[HBr]	[H₂]	[Br₂]
I	0.500	–	–
C	–2x	+x	+x
E	0.500–2x	+x	+x

The problem tell us that $[Br_2] = 0.0955$ M = x at equilibrium. Using the ICE table as a guide we see that the equilibrium concentrations are; $[H_2] = [Br_2] = 0.0955$ M and $[HBr] = 0.500 - 2(0.0955) = 0.309$ M.

$$K_c = \frac{[H_2][Br_2]}{[HBr]^2} = \frac{(0.0955)(0.0955)}{(0.309)^2} = 0.0955$$

15.72 According to the problem, the concentration of NO_2 increases in the course of this reaction. This means our ICE table will look like the following:

	[NO₂]	[NO]	[N₂O]	[O₂]
I	0.0560	0.294	0.184	0.377
C	+x	+x	–x	–x
E	0.0560+x	0.294+x	0.184–x	0.377–x

The problem tell us that $[NO_2] = 0.118$ M = 0.0560+x at equilibrium. Solving we get; x = 0.062 M. Using the ICE table as a guide we see that the equilibrium concentrations are; $[NO] = 0.356$ M, $[N_2O] = 0.122$ M and $[O_2] = 0.315$ M.

$$K_c = \frac{[N_2O][O_2]}{[NO_2][NO]} = \frac{(0.122)(0.315)}{(0.118)(0.356)} = 0.915$$

15.74 The initial concentrations are each 0.300 mol/2.00 L = 0.150 M.

	[SO₃]	[NO]	[NO₂]	[SO₂]
I	–	–	0.300	0.300
C	+x	+x	–x	–x
E	+x	+x	0.150–x	0.150–x

Substituting the above values for equilibrium concentrations into the mass action expression gives:

$$K_c = \frac{[NO_2][SO_2]}{[SO_3][NO]} = \frac{(0.150-x)(0.150-x)}{(x)(x)} = 0.500$$

Taking the square root of both sides of this equation gives: $0.707 = (0.150-x)/x$. Solving for x we have: $1.707(x) = 0.150$ ∴ $x = 0.0878$ mol/L = $[NO] = [SO_3]$, $[NO_2] = [SO_2] = 0.150 - x = 0.062$ mol/L

Although these two systems reach different equilibrium positions, the equilibrium concentrations of the four substances in each experiment give the same value for the equilibrium constant when substituted into the mass action expression.

An alternative method for solving problems where the reaction proceeds from right to left, such as this problem, is to reverse the reaction written and use this "new" reaction as the basis for the problem. In doing this, the mass action expression and the equilibrium constant must change to reflect the "new" reaction. As outlined in Section 15.3, "when the direction of an equation is reversed, the new equilibrium constant is the reciprocal of the original" and the "new" mass action expression is also the reciprocal of the original. Try solving this problem using this method. Your answer should, of course, remain the same as stated above.

15.76

	[HCl]	[H₂]	[Cl₂]
I	0.0500	–	–
C	–2x	+x	+x
E	0.0500–2x	+x	+x

Substituting the above values for equilibrium concentrations into the mass action expression gives:

$$K_c = \frac{[H_2][Cl_2]}{[HCl]^2} = \frac{(x)(x)}{(0.0500-2x)^2} = 3.2 \times 10^{-34}$$

Because K_c is so exceedingly small, we can make the simplifying assumption that x is also small enough to make $(0.0500 - 2x) \approx 0.0500$. Thus we have: $3.2 \times 10^{-34} = (x)^2/(0.0500)^2$

Taking the square root of both sides, and solving for the value of x gives:

$x = 8.9 \times 10^{-19}$ M $= [H_2] = [Cl_2]$

$[HCl] = (0.0500 - x) \approx 0.0500$ mol/L

15.78 The initial concentrations are :

0.200 mol/4.00 L = 0.0500 M N_2O

0.400 mol/4.00 L = 0.100 M NO_2

	$[N_2O]$	$[NO_2]$	$[NO]$
I	0.0500	0.100	–
C	–x	–x	+3x
E	0.0500–x	0.100–x	+3x

Substituting the above values for equilibrium concentrations into the mass action expression gives:

$$K_c = \frac{[NO]^3}{[N_2O][NO_2]} = \frac{(3x)^3}{(0.0500 - x)(0.100 - x)} = 1.4 \times 10^{-10}$$

Make the usual assumption that x is small and we can simplify this expression to:

$x^3 = 2.6 \times 10^{-14}$ and $x = 3.0 \times 10^{-5}$ M

$[NO] = 3x = 8.9 \times 10^{-5}$ M. (Note: This value was not rounded until the very end of the calculation, i.e., the unrounded value determined for x was multiplied by 3.)

15.80 The initial concentrations are: 0.200 mol/5.00 L = 0.0400 M. Since the ion product constant for this system is greater than K_c, this reaction will proceed from right to left as written.

	$[NO_2]$	$[NO]$	$[N_2O]$	$[O_2]$
I	0.0400	0.0400	0.0400	0.0400
C	+x	+x	–x	–x
E	0.0400+x	0.0400+x	0.0400–x	0.0400–x

Substituting the above values for equilibrium concentrations into the mass action expression gives:

$$K_c = \frac{[N_2O][O_2]}{[NO_2][NO]} = \frac{(0.0400 - x)(0.0400 - x)}{(0.0400 + x)(0.0400 + x)} = 0.914$$

Taking the square root of both sides of the equation:

$$0.956 = \frac{(0.0400 - x)}{(0.0400 + x)}$$

$0.0382 + 0.956(x) = 0.0400 - x$ and $x = 9.20 \times 10^{-4}$ M

$[N_2O] = [O_2] = 0.0400 - 0.000920 = 0.0391$ M
$[NO_2] = [NO] = 0.0400 + 0.000920 = 0.0409$ M

An alternative method for solving this problem is described in Review Exercise number 15.74.

15.82   All concentrations are a factor of ten less than the stated value due to the container size. Since $Q < K_c$ for this reaction, the reaction proceeds as written, i.e., from left to right.

	$[SO_2]$	$[NO_2]$	$[NO]$	$[SO_3]$
I	0.0100	0.00600	0.00800	0.0120
C	$-x$	$-x$	$+x$	$+x$
E	$0.0100-x$	$0.00600-x$	$0.00800+x$	$0.0120+x$

Substituting the above values for equilibrium concentrations into the mass action expression gives:

$$K_c = \frac{[NO][SO_3]}{[SO_2][NO_2]} = \frac{(0.00800 + x)(0.0120 + x)}{(0.0100 - x)(0.00600 - x)} = 85.0$$

$0.00510 - 1.36(x) + 85.0(x)^2 = 9.60 \times 10^{-5} + 0.0200(x) + (x)^2$
$84.0(x)^2 - 1.38(x) + 0.00500 = 0$

The quadratic equation is used, with the following values: $a = 84.0$; $b = -1.38$; $c = 0.00500$. Upon solving for x, we find that only the negative root is sensible: $x = 0.00542$ M.

$[SO_2] = 0.0100 - 0.00542 = 0.0046$ M
$[NO_2] = 0.00600 - 0.00542 = 0.00058$ M
$[NO] = 0.00800 + 0.00542 = 0.01342$ M
$[SO_3] = 0.0120 + 0.00542 = 0.0174$ M

15.84

	[SO₃]	[NO]	[NO₂]	[SO₂]
I	0.0500	0.100	–	–
C	–x	–x	+x	+x
E	0.0500–x	0.100–x	+x	+x

Substituting the above values for equilibrium concentrations into the mass action expression gives:

$$K_c = \frac{[NO_2][SO_2]}{[SO_3][NO]} = \frac{(x)(x)}{(0.0500-x)(0.100-x)} = 0.500$$

Since the equilibrium constant is not much larger than either of the values 0.0500 or 0.100, we cannot neglect the size of x in the above expression. A simplifying assumption is not therefore possible, and we must solve for the value of x using the quadratic equation. Multiplying out the above denominator, collecting like terms, and putting the result into the standard quadratic form gives:

$$0.500x^2 + (7.50 \times 10^{-2})x - (2.50 \times 10^{-3}) = 0$$

$$x = \frac{-7.50 \times 10^{-2} \pm \sqrt{(-7.50 \times 10^{-2})^2 - 4(0.500)(-2.50 \times 10^{-3})}}{2(0.500)} = 0.0281 \text{ M,}$$

using the (+) root. So, [NO₂] = [SO₂] = 0.0281 M

15.86 If the problem is worked in the usual fashion, the value for x is found to be essentially zero due to the large value for the equilibrium constant:

	[HCHO₂]	[CO]	[H₂O]
I	0.200	–	–
C	–x	+x	+x
E	0.200–x	+x	+x

Substituting the above values for equilibrium concentrations into the mass action expression gives:

$$K_c = \frac{[CO][H_2O]}{[HCHO_2]} = \frac{(x)(x)}{(0.200-x)} = 4.3 \times 10^4$$

Rearranging to a form suitable for the quadratic equation gives:

$(x)^2 + 4.3 \times 10^4(x) - 8.6 \times 10^3 = 0$. Therefore: $a = 1$; $b = 4.3 \times 10^4$; $c = -8.6 \times 10^3$

Use of the quadratic equation gives a value of zero for x. Since the value for x may be a very small number, but may not be exactly zero, a different solution to the problem must be sought. The best approach is to "allow" the original initial conditions to become adjusted completely to the other side of the equilibrium. In this example, we "allow" all of the original $HCHO_2$ to react, giving itself a new initial concentration of 0 M, and giving CO and H_2O each new initial concentrations of 0.200 M. The new equilibrium problem can then be solved in the usual manner, this time giving an equilibrium that correctly favors products.

	$[HCHO_2]$	$[CO]$	$[H_2O]$
I	–	0.200	0.200
C	$+x$	$-x$	$-x$
E	$+x$	$0.200-x$	$0.200-x$

Substituting the above values for the equilibrium concentrations into the mass action expression gives:

$$K_c = \frac{[CO][H_2O]}{[HCHO_2]} = \frac{(0.200 - x)(0.200 - x)}{(x)} = 4.3 \times 10^4$$

The simplifying assumption can be made that x is very small compared to the number 0.200. Solving for x gives: $4.3 \times 10^4(x) = 0.0400$ ∴ $x = [HCHO_2] = 9.3 \times 10^{-7}$ M
$[CO] = [H_2O] = 0.200 - x = 0.200$ M

15.88 (a)   Equilibrium is unaffected by the addition of a solid.
     (b)   Equilibrium is unaffected by the removal of a solid.
     (c)   The equilibrium will shift to the left.
     (d)   The equilibrium will shift to the right.

CHAPTER SIXTEEN
Review Exercises

16.3 It should be stated from the outset that water at this temperature is neutral by definition, since $[H^+] = [OH^-]$. In other words, the self–ionization of water still occurs on a one–to–one mole basis: $H_2O \rightleftharpoons H^+ + OH^-$.

$K_w = 2.4 \times 10^{-14} = [H^+] \times [OH^-]$

Since $[H^+] = [OH^-]$, we can rewrite the above relationship:

$2.4 \times 10^{-14} = ([H^+])^2$, $\therefore [H^+] = [OH^-] = 1.5 \times 10^{-7}$ M

$pH = -\log[H^+] = -\log(1.5 \times 10^{-7}) = 6.82$

$pOH = -\log[OH^-] = -\log(1.5 \times 10^{-7}) = 6.82$

$pK_w = pH + pOH = 6.82 + 6.82 = 13.64$

Alternatively, for the last calculation we can write:

$pK_w = -\log(K_w) = -\log(2.4 \times 10^{-14}) = 13.62$

Water is neutral at this temperature because the concentration of the hydrogen ion is the same as the concentration of the hydroxide ion.

16.5 At 25 °C, $K_w = 1.0 \times 10^{-14} = [H^+] \times [OH^-]$. Let $x = [H^+]$, for each of the following:

  (a) $x(0.0024) = 1.0 \times 10^{-14}$
       $[H^+] = (1.0 \times 10^{-14}) \div (0.0024) = 4.2 \times 10^{-12}$ M

  (b) $x(1.4 \times 10^{-5}) = 1.0 \times 10^{-14}$
       $[H^+] = (1.0 \times 10^{-14}) \div (1.4 \times 10^{-5}) = 7.1 \times 10^{-10}$ M

  (c) $x(5.6 \times 10^{-9}) = 1.0 \times 10^{-14}$
       $[H^+] = (1.0 \times 10^{-14}) \div (5.6 \times 10^{-9}) = 1.8 \times 10^{-6}$ M

  (d) $x(4.2 \times 10^{-13}) = 1.0 \times 10^{-14}$
       $[H^+] = (1.0 \times 10^{-14}) \div (4.2 \times 10^{-13}) = 2.4 \times 10^{-2}$ M

16.6 At 25 °C, $K_w = 1.0 \times 10^{-14} = [H^+] \times [OH^-]$. Let $x = [OH^-]$, for each of the following:

  (a) $(3.5 \times 10^{-8})x = 1.0 \times 10^{-14}$
       $[OH^-] = (1.0 \times 10^{-14}) \div (3.5 \times 10^{-8}) = 2.9 \times 10^{-7}$ M

(b) $(0.0065)x = 1.0 \times 10^{-14}$

$[OH^-] = (1.0 \times 10^{-14}) \div (0.0065) = 1.5 \times 10^{-12}$ M

(c) $(2.5 \times 10^{-13})x = 1.0 \times 10^{-14}$

$[OH^-] = (1.0 \times 10^{-14}) \div (2.5 \times 10^{-13}) = 4.0 \times 10^{-2}$ M

(d) $(7.5 \times 10^{-5})x = 1.0 \times 10^{-14}$

$[OH^-] = (1.0 \times 10^{-14}) \div (7.5 \times 10^{-5}) = 1.3 \times 10^{-10}$ M

16.8 $pH = -\log[H^+] = -\log(1.4 \times 10^{-5}) = 4.85$

16.10 $[H^+] = 10^{-pH}$ and $[OH^-] = 10^{-pOH}$
At 25 °C, $pH + pOH = 14.00$

(a) $[H^+] = 10^{-pH} = 10^{-3.14} = 7.2 \times 10^{-4}$ M

$pOH = 14.00 - pH = 14.00 - 3.14 = 10.86$

$[OH^-] = 10^{-pOH} = 10^{-10.86} = 1.4 \times 10^{-11}$ M

(b) $[H^+] = 10^{-pH} = 10^{-2.78} = 1.7 \times 10^{-3}$ M

$pOH = 14.00 - pH = 14.00 - 2.78 = 11.22$

$[OH^-] = 10^{-pOH} = 10^{-11.22} = 6.0 \times 10^{-12}$ M

(c) $[H^+] = 10^{-pH} = 10^{-9.25} = 5.6 \times 10^{-10}$ M

$pOH = 14.00 - pH = 14.00 - 9.25 = 4.75$

$[OH^-] = 10^{-pOH} = 10^{-4.75} = 1.8 \times 10^{-5}$ M

(d) $[H^+] = 10^{-pH} = 10^{-13.24} = 5.8 \times 10^{-14}$ M

$pOH = 14.00 - pH = 14.00 - 13.24 = 0.76$

$[OH^-] = 10^{-pOH} = 10^{-0.76} = 1.7 \times 10^{-1}$ M

(e) $[H^+] = 10^{-pH} = 10^{-5.70} = 2.0 \times 10^{-6}$ M

$pOH = 14.00 - pH = 14.00 - 5.70 = 8.30$

$[OH^-] = 10^{-pOH} = 10^{-8.30} = 5.0 \times 10^{-9}$ M

16.12   HCl is a strong acid so $[H^+] = [HCl] = 0.010$ M

$pH = -\log[H^+] = -\log(0.010) = 2.00$

16.14 $\quad M\,OH^- = \dfrac{\text{\# moles OH}^-}{\text{\# L solution}} = \left(\dfrac{6.0\text{ g NaOH}}{1.00\text{ L solution}}\right)\left(\dfrac{1\text{ mole NaOH}}{40.0\text{ g NaOH}}\right)\left(\dfrac{1\text{ mole OH}^-}{1\text{ mole NaOH}}\right)$

$\quad\quad = 0.15\,M\,OH^-$

$\quad pOH = -\log[OH^-] = -\log(0.15) = 0.83$
$\quad pH = 14.00 - pOH = 14.00 - 0.83 = 13.17$

16.16 $\quad pOH = 14.00 - pH = 14.00 - 11.60 = 2.40$

$\quad [OH^-] = 10^{-pOH} = 10^{-2.40} = 4.0 \times 10^{-3}\,M$

$\quad [Ca(OH)_2] = \left(\dfrac{4.0 \times 10^{-3}\text{ mol OH}^-}{1\text{ L solution}}\right)\left(\dfrac{1\text{ mol Ca(OH)}_2}{2\text{ mol OH}^-}\right)$

$\quad\quad = 2.0 \times 10^{-3}\,M\,Ca(OH)_2$

16.18   (a) $HNO_2 \rightleftharpoons H^+ + NO_2^-$

       (b) $H_3PO_4 \rightleftharpoons H^+ + H_2PO_4^-$

       (c) $HAsO_4^{2-} \rightleftharpoons H^+ + AsO_4^{3-}$

       (d) $(CH_3)_3NH^+ \rightleftharpoons H^+ + (CH_3)_3N$

16.20   (a) $(CH_3)_3N + H_2O \rightleftharpoons (CH_3)_3NH^+ + OH^-$

       (b) $AsO_4^{3-} + H_2O \rightleftharpoons HAsO_4^{2-} + OH^-$

       (c) $NO_2^- + H_2O \rightleftharpoons HNO_2 + OH^-$

       (d) $(CH_3)_2N_2H_2 + H_2O \rightleftharpoons (CH_3)_2N_2H_3^+ + OH^-$

16.24   In general, there is an inverse relationship between the strength of an acid and its conjugate base; the stronger the acid, the weaker its conjugate base. Since HCN has the larger value for pK_a, we know that it is a weaker acid than HF. Accordingly, CN^- is a stronger base than F^-.

16.26   At 25 °C, $K_a \times K_b = K_w$

$\quad K_a = K_w/K_b = 1.0 \times 10^{-14} \div 1.0 \times 10^{-10} = 1.0 \times 10^{-4}$

16.28 $K_a \times K_b = K_w$

$\quad K_a = K_w/K_b = 1.0 \times 10^{-14} \div 4.4 \times 10^{-4} = 2.3 \times 10^{-11}$

16.30 (a)   The conjugate base is IO_3^-

$pK_b = 14.00 - pK_a = 14.00 - 0.77 = 13.23$

$K_b = 10^{-pK_b} = 10^{-13.23} = 5.9 \times 10^{-14}$

(b) IO_3^- is a weaker base than an acetate anion, because its K_b value is smaller than that of an acetate anion.

16.32 $HC_2H_2ClO_2 \rightleftharpoons H^+ + C_2H_2ClO_2^-$

$$K_a = \frac{[H^+][C_2H_2ClO_2^-]}{[HC_2H_2ClO_2]}$$

	$[HC_2H_2ClO_2]$	$[H^+]$	$[C_2H_2ClO_2^-]$
I	0.10	–	–
C	–x	+x	+x
E	0.10–x	+x	+x

We know that the acid is 11.0% ionized so x = 0.10 M × 0.11 = 0.011 M. Therefore, our equilibrium concentrations are:

$[H^+] = [C_2H_2ClO_2^-] = 0.011$ M, and $[HC_2H_2ClO_2] = 0.10$ M – 0.011 M = 0.09 M.

Substituting these values into the mass action expression gives:

$$K_a = \frac{(0.011)(0.011)}{0.09} = 1 \times 10^{-3}$$

$pK_a = -\log(K_a) = -\log(1.0 \times 10^{-3}) = 3.0$

16.34 $HONH_2 + H_2O \rightleftharpoons HONH_3^+ + OH^-$

$$K_b = \frac{[HONH_3^+][OH^-]}{[HONH_2]}$$

$pOH = 14.00 - pH = 14.00 - 10.12 = 3.88$

$[OH^-] = 10^{-3.88} = 1.3 \times 10^{-4}$ M

Chapter Sixteen

In the equilibrium analysis, the value of x is, therefore, equal to 1.3×10^{-4} M:

	$[HONH_2]$	$[HONH_3^+]$	$[OH^-]$
I	0.15	–	–
C	–x	+x	+x
E	0.15–x	+x	+x

Therefore, our equilibrium concentrations are $[HONH_3^+] = [OH^-] = 1.3 \times 10^{-4}$ M, and $[HONH_2] = 0.15$ M $- 1.3 \times 10^{-4}$ M $= 0.15$ M.

Substituting these values into the mass action expression gives:

$$K_b = \frac{(1.3 \times 10^{-4})(1.3 \times 10^{-4})}{0.15} = 1.1 \times 10^{-7}$$

$pK_b = -\log(K_b) = -\log(1.1 \times 10^{-7}) = 6.96$

16.36 We can generalize the acid dissociation as $HA \rightleftharpoons H^+ + A^-$.

$$K_a = \frac{[H^+][A^-]}{[HA]} = \frac{(x)(x)}{(0.125 - x)} = 3.2 \times 10^{-3}$$

For this problem, we must use the method of successive approximations (or the quadratic equation), because the value of K_a is fairly large. We start by assuming that x is small compared to 0.125 (an assumption which we anticipate will not be altogether proper) and determine that x = 0.020. As we can see, the value for x is not insignificant when compared to the initial concentration of pyruvic acid. Substituting this value of x into the denominator of the mass action expression and solving for x in the numerator gives us our second iteration:

$3.2 \times 10^{-3} = x^2/(0.125 - 0.020)$; x = 0.018.

A third iteration is in order: $3.2 \times 10^{-3} = x^2/(0.125 - 0.018)$; x = 0.019

A fourth iteration gives x = 0.018 M = $[H^+]$; pH = $-\log(0.018)$ = 1.74
Using the method of successive approximations, we continue iterating until the calculated value does not change.

143

16.38 $H_2O_2 \rightleftharpoons H^+ + HO_2^-$

$$K_a = \frac{[H^+][HO_2^-]}{[H_2O_2]} = 1.8 \times 10^{-12}$$

	$[H_2O_2]$	$[H^+]$	$[HO_2^-]$
I	1.0	—	—
C	–x	+x	+x
E	1.0–x	+x	+x

Substituting these values into the mass action expression gives:

$$K_a = \frac{(x)(x)}{1.0-x} = 1.8 \times 10^{-12}$$

Assuming x << 1.0 we determine that x = 1.3×10^{-6} M = $[H^+]$
pH = $-\log[H^+]$ = 5.89

16.40 $K_b = 10^{-pK_b} = 10^{-5.79} = 1.6 \times 10^{-6}$

$Cod + H_2O \rightleftharpoons HCod^+ + OH^-$

$$K_b = \frac{[HCod^+][OH^-]}{[Cod]} = 1.6 \times 10^{-6}$$

	$[Cod]$	$[HCod^+]$	$[OH^-]$
I	0.020	—	—
C	–x	+x	+x
E	0.020–x	+x	+x

Substituting these values into the mass action expression gives:

$$K_b = \frac{(x)(x)}{0.020-x} = 1.6 \times 10^{-6}$$

If we assume that x << 0.020 we get; $x^2 = 3.2 \times 10^{-8}$,
∴ x = 1.8×10^{-4} M = $[OH^-]$
pOH = $-\log[OH^-]$ = $-\log(1.8 \times 10^{-4})$ = 3.74
pH = 14.00 – pOH = 14.00 – 3.74 = 10.26

16.42 $[H^+] = 10^{-pH} = 10^{-2.54} = 2.9 \times 10^{-3}$ M

$$HC_2H_3O_2 \rightleftharpoons H^+ + C_2H_3O_2^-$$

$$K_a = \frac{[H^+][C_2H_3O_2^-]}{[HC_2H_3O_2]} = 1.8 \times 10^{-5}$$

	$[HC_2H_3O_2]$	$[H^+]$	$[C_2H_3O_2^-]$
I	Z	–	–
C	–x	+x	+x
E	Z–x	+x	+x

Substituting these values into the mass action expression gives:

$$K_a = \frac{(x)(x)}{Z-x} = 1.8 \times 10^{-5}$$

Assuming $x \ll Z$ and knowing that $x = 2.9 \times 10^{-3}$ M, we can solve for Z and find Z = 0.47. The initial concentration of $HC_2H_3O_2$ is 0.47 M.

16.44 The oxalate ion hydrolyzes in water, to give a basic solution, according to the equation:
$$C_2O_4^{2-} + H_2O \rightleftharpoons HC_2O_4^- + OH^-.$$

16.49 NaCN will be basic in solution since CN^- is a basic ion and Na^+ is a neutral ion.

$$CN^- + H_2O \rightleftharpoons HCN + OH^-$$

For HCN, $K_a = 6.2 \times 10^{-10}$, we need K_b for CN^-;

$$K_b = K_w/K_a = (1.0 \times 10^{-14}) \div (6.2 \times 10^{-10}) = 1.6 \times 10^{-5}$$

$$K_b = \frac{[HCN][OH^-]}{[CN^-]} = 1.6 \times 10^{-5}$$

	$[CN^-]$	$[HCN]$	$[OH^-]$
I	0.20	–	–
C	–x	+x	+x
E	0.20–x	+x	+x

Substituting these values into the mass action expression gives:

$$K_b = \frac{(x)(x)}{0.20 - x} = 1.6 \times 10^{-5}$$

Assuming that $x \ll 0.20$ we can solve for x an determine;

$x = 1.8 \times 10^{-3}$ M $= [OH^-]$

$pOH = -\log[OH^-] = -\log(1.8 \times 10^{-3}) = 2.74$

$pH = 14.00 - pOH = 14.00 - 2.74 = 11.26$

16.51　A solution of CH_3NH_3Cl will be acidic since the Cl^- ion is neutral and the $CH_3NH_3^+$ ion is acidic.

$$CH_3NH_3^+ \rightleftharpoons H^+ + CH_3NH_2$$

For CH_3NH_2, $K_b = 4.4 \times 10^{-4}$. We need K_a for $CH_3NH_3^+$;

$K_a = K_w/K_b = (1.0 \times 10^{-14}) \div (4.4 \times 10^{-4}) = 2.3 \times 10^{-11}$

$$K_a = \frac{[H^+][CH_3NH_2]}{[CH_3NH_3^+]} = 2.3 \times 10^{-11}$$

	$[CH_3NH_3^+]$	$[H^+]$	$[CH_3NH_2]$
I	0.15	–	–
C	–x	+x	+x
E	0.15–x	+x	+x

Substituting these values into the mass action expression gives:

$$K_a = \frac{(x)(x)}{0.15 - x} = 2.3 \times 10^{-11}$$

Assuming that $x \ll 0.15$ we can solve for x an determine;

$x = 1.9 \times 10^{-6}$ M $= [H_3O^+]$

$pH = -\log[H_3O^+] = -\log(1.9 \times 10^{-6}) = 5.72$

16.53　The only dissociation that we need to consider is: $NH_4^+ \rightleftharpoons NH_3 + H^+$,

if pH = 5.16, then $[H^+] = 10^{-pH} = 10^{-5.16} = 6.9 \times 10^{-6}$ M

This is also the required value for $[NH_3]$, since they are formed in a one-to-one mole ratio.

$K_a = [H^+][NH_3]/[NH_4^+] = 5.7 \times 10^{-10}$ and we arrive at:

$5.7 \times 10^{-10} = (6.9 \times 10^{-6})(6.9 \times 10^{-6})/[NH_4^+]$

$[NH_4^+] = 8.4 \times 10^{-2}$ M

8.4×10^{-2} mol/L $\times$ 98.0 g/mol = 8.2 g NH_4Br are needed per liter.

16.55 The reaction for this problem is:

H–Mor$^+$ $\rightleftharpoons$ H$^+$ + Mor

We know that $pK_a + pK_b = pK_w = 14.00$.

So, $pK_a = 14.00 - pK_b = 14.00 - 6.13 = 7.87$, and $K_a = 10^{-pK_a} = 1.3 \times 10^{-8}$.

$$K_a = \frac{[H^+][Mor]}{[H - Mor^+]} = 1.3 \times 10^{-8}$$

	[H–Mor$^+$]	[H$^+$]	[Mor]
I	0.20	–	–
C	–x	+x	+x
E	0.20–x	+x	+x

Substituting these values into the mass action expression gives:

$$K_a = \frac{(x)(x)}{0.20 - x} = 1.3 \times 10^{-8}$$

Assuming that x << 0.20 we can solve for x an determine;

$x = 5.1 \times 10^{-5}$ M = [H$^+$]

$pH = -\log[H^+] = -\log(5.1 \times 10^{-5}) = 4.29$

16.58 HF $\rightleftharpoons$ H$^+$ + F$^-$

$$K_a = \frac{[H^+][F^-]}{[HF]} = 6.8 \times 10^{-4}$$

	[HF]	[H$^+$]	[F$^-$]
I	0.15	–	–
C	–x	+x	+x
E	0.15–x	+x	+x

Substituting the above values for equilibrium concentrations into the mass action expression and assuming that x << 0.15 gives:

$$K_a = \frac{(x)(x)}{0.15} = 6.8 \times 10^{-4}$$

$x^2 = 1.0 \times 10^{-4}, \therefore x = 1.0 \times 10^{-2} \, M = [H^+]$

$pH = -\log[H^+] = -\log(0.010) = 2.00$

% ionization $= 0.010/0.15 \times 100 = 6.7\%$

Since the % ionization is > 5%, we can not make the simplifying assumption. Consequently, we need to solve for x using a quadratic equation. The mass action expression is:

$$K_a = \frac{(x)(x)}{0.15 - x} = 3.0 \times 10^{-9}$$

We may rearrange this expression to obtain the following quadratic equation:

$$x^2 + 6.8 \times 10^{-4}x - 1.0 \times 10^{-4} = 0$$

where a = 1, b = 6.8 $\times 10^{-4}$, and c = 1.0 $\times 10^{-4}$. If we substitute these values into the quadratic equation (see Example 16.16) and take the positive root we get x = 0.0097 M.

$[H^+] = x = 0.0097 \, M.$

$pH = -\log[H^+] = -\log(0.0097) = 2.01$

% ionization $= 0.0097/0.15 \times 100 = 6.5\%$

In this problem, the change in pH between assuming x to be small and not assuming x to be small is very slight.

16.60 HCl is a strong acid and, consequently, solutions of HCl should be acidic. If we assume that the HCl ionizes completely, we determine that the $[H^+] = 1.0 \times 10^{-7}$ and the pH is 7.00. This is a neutral solution. For this problem, we must also account for the autoionization of H_2O. It too contributes to the $[H^+]$. The total $[H^+]$ will be the sum of these and equals $1.0 \times 10^{-7} + 1.0 \times 10^{-7} = 2.0 \times 10^{-7} \, M$. The pH is the $-\log(2.0 \times 10^{-7})$ = 6.70. The final pH is acidic as we expected.

16.62 $K_a = 10^{-pK_a} = 10^{-4.92} = 1.2 \times 10^{-5}$

$H{-}Paba \rightleftharpoons H^+ + Paba^-$

$$K_a = \frac{\left[H^+\right]\left[Paba^-\right]}{\left[H - Paba\right]} = 1.2 \times 10^{-5}$$

	[H–Paba]	[H$^+$]	[Paba$^-$]
I	0.030	–	–
C	–x	+x	+x
E	0.030–x	+x	+x

Substituting the above values for equilibrium concentrations into the mass action expression and assuming that x << 0.030 gives:

$$K_a = \frac{[x][x]}{[0.030]} = 1.2 \times 10^{-5}$$

$$x^2 = 3.6 \times 10^{-7}, \quad \therefore x = 6.0 \times 10^{-4} \text{ M} = [\text{H}^+]$$

$$\text{pH} = -\log[\text{H}^+] = -\log(6.0 \times 10^{-4}) = 3.22$$

16.66 $HC_2H_3O_2 \rightleftharpoons H^+ + C_2H_3O_2^-$

$$K_a = \frac{[H^+][C_2H_3O_2^-]}{[HC_2H_3O_2]} = 1.8 \times 10^{-5}$$

	[HC$_2$H$_3$O$_2$]	[H$^+$]	[C$_2$H$_3$O$_2^-$]
I	0.15	–	0.25
C	–x	+x	+x
E	0.15–x	+x	0.25+x

Substituting these values into the mass action expression gives:

$$K_a = \frac{(x)(0.25 + x)}{0.15 - x} = 1.8 \times 10^{-5}$$

Assume that x << 0.15 M and x << 0.25 M, then;

$$x \times \left(\frac{0.25}{0.15}\right) \approx 1.8 \times 10^{-5}$$

$$x \approx \left(\frac{0.15}{0.25}\right) \times 1.8 \times 10^{-5}$$

$$x \approx 1.1 \times 10^{-5} \text{ M} = \left[H^+\right]$$

$$\text{pH} = -\log\left[H^+\right] = 4.97$$

16.68 The initial pH of the buffer is 4.97 as determined in Exercise 16.66. The added acid, 0.050 mol, will react with the acetate ion present in the buffer solution. Assume the added acid reacts completely. For each mole of acid added, one mole of $C_2H_3O_2^-$ is converted to $HC_2H_3O_2$. Since 0.050 mol of acid is added;

$$[HC_2H_3O_2]_{final} = (0.15 + 0.050)\ M = 0.20\ M$$

$$[C_2H_3O_2^-]_{final} = (0.25 - 0.050)\ M = 0.20\ M$$

Now, substitute these values into the mass action expression to calculate the final $[H^+]$ in solution;

$$\frac{\left[H^+\right](0.20)}{(0.20)} = 1.8 \times 10^{-5}$$

$[H^+] = 1.8 \times 10^{-5}\ mol\ L^{-1}$ and the pH = 4.74

The pH of the solution changes by $4.74 - 4.93 = -0.23$ pH units upon addition of the acid.

16.70 The equilibrium we will consider in this problem is: $NH_3 + H_2O \rightleftharpoons NH_4^+ + OH^-$

a) $$K_b = \frac{\left[NH_4^+\right]\left[OH^-\right]}{[NH_3]} = 1.8 \times 10^{-5}$$

$$= \frac{(0.14)\left[OH^-\right]}{0.25} = 1.8 \times 10^{-5}$$

$[OH^-] = 3.2 \times 10^{-5}\ M$

$pOH = -\log[OH^-] = -\log(3.2 \times 10^{-5}) = 4.49$

$pH = 14.00 - pOH = 14.00 - 4.49 = 9.51$

(b) Now we consider the equilibrium; $NH_4^+ \rightleftharpoons H^+ + NH_3$

We know that $K_a K_b = K_w$, so $K_a = K_w/K_b = 5.6 \times 10^{-10}$.

$$K_a = \frac{\left[H^+\right]\left[NH_3\right]}{\left[NH_4^+\right]} = 5.6 \times 10^{-10}$$

$$= \frac{\left[H^+\right](0.25)}{0.14} = 5.6 \times 10^{-10}$$

$$[H^+] = 3.1 \times 10^{-10} \text{ M and pH} = 9.50$$

Notice that the two values calculated are essentially the same.

16.72 The initial pH is 9.51 as calculate in Exercise 16.70. First we will determine the number of moles of OH^- added:

$$\# \text{ mol } OH^- = (75 \text{ mL KOH})\left(\frac{0.10 \text{ mol KOH}}{1000 \text{ mL KOH}}\right)\left(\frac{1 \text{ mol } OH^-}{1 \text{ mol KOH}}\right)$$

$$= 7.5 \times 10^{-3} \text{ mol } OH^-$$

Then determine the number of moles of NH_4^+ and NH_3 initially present:

$$\# \text{ mol } NH_4^+ = (200 \text{ mL solution})\left(\frac{0.14 \text{ mol } NH_4^+}{1000 \text{ mL solution}}\right)$$

$$= 2.8 \times 10^{-2} \text{ mol } NH_4^+$$

$$\# \text{ mol } NH_3 = (200 \text{ mL solution})\left(\frac{0.25 \text{ mol } NH_3}{1000 \text{ mL solution}}\right)$$

$$= 5.0 \times 10^{-2} \text{ mol } NH_3$$

For every mole of OH^- added, one mol of NH_4^+ will be changed to one mol of NH_3. Since we added 7.5×10^{-3} mol OH^-;

$$\left[NH_4^+\right]_{final} = \frac{(0.028 \text{ mol} - 0.0075 \text{ mol})}{0.275 \text{ L}} = 0.075 \text{ M}$$

$$\left[NH_3\right]_{final} = \frac{(0.050 \text{ mol} + 0.0075 \text{ mol})}{0.275 \text{ L}} = 0.21 \text{ M}$$

Note: the volume of the solution changes as a result of the addition of KOH.

Using these new concentrations, we can calculate a new pH;

$$K_b = \frac{\left[NH_4^+\right]\left[OH^-\right]}{\left[NH_3\right]} = \frac{(0.075)\left[OH^-\right]}{0.21} = 1.8 \times 10^{-5}$$

$[OH^-] = 5.0 \times 10^{-5}$ M, the pOH = 4.30 and the pH = 9.70.

As expected, when a base is added, the pH increases. In this problem, the pH increases by 0.19 pH units, from 9.51 to 9.70.

16.74 $pH = pK_a + \log \dfrac{[\text{anion}]}{[\text{acid}]} = pK_a + \log \dfrac{[A^-]}{[HA]}$

$3.80 = 3.74 + \log([NaCHO_2]/[HCHO_2])$
$[NaCHO_2]/[HCHO_2] = 1.1$
$[NaCHO_2] = 1.1 \times [HCHO_2] = 1.1 \times 0.12 = 0.13\ M$

Thus to the 1 L of formic acid solution we add: 0.13 mol $NaCHO_2 \times 68.0$ g/mol = 8.8 g $NaCHO_2$.

16.76 $pOH = pK_b + \log \dfrac{[\text{cation}]}{[\text{base}]}$

Now pOH = 14.00 − pH = 14.00 − 10.00 = 4.00
$4.00 = 4.74 + \log([NH_4^+]/[NH_3])$
$\log([NH_4^+]/[NH_3]) = -0.74$
Taking the antilog of both sides of this equation gives:
$[NH_4^+]/[NH_3] = 10^{-0.74} = 0.18$
Thus $[NH_4^+] = 0.18 \times [NH_3] = 0.18 \times 0.20\ M = 3.6 \times 10^{-2}\ M = [NH_4Cl]$

Finally, 0.500 L of buffer solution would require:
$0.500\ L \times 3.6 \times 10^{-2}\ mol/L \times 53.5\ g/mol = 0.96\ g\ NH_4Cl$

16.78   The equilibrium is; $HC_2H_3O_2 \rightleftharpoons H^+ + C_2H_3O_2^-$

$K_a = \dfrac{[H^+][C_2H_3O_2^-]}{[HC_2H_3O_2]} = 1.8 \times 10^{-5}$

The initial pH is; $\dfrac{[H^+](0.110)}{0.100} = 1.8 \times 10^{-5}$, $[H^+] = 1.64 \times 10^{-5}\ M$ and pH = 4.786. In this calculation we are able to use either the molar concentration or the number of moles since the volume is constant in this portion of the problem.

Chapter Sixteen

In order to calculate the change in pH, we need to determine the concentrations of $HC_2H_3O_2$ and $C_2H_3O_2^-$ after the complete reaction of the added acid. One mole of $C_2H_3O_2^-$ will be consumed for every mole of acid added and one mole of $HC_2H_3O_2$ will be produced. The number of moles of acid added is;

$$\text{\# mol } H^+ = (25.00 \text{ mL HCl})\left(\frac{0.100 \text{ mol HCl}}{1000 \text{ mL HCl}}\right)\left(\frac{1 \text{ mol } H^+}{1 \text{ mol HCl}}\right)$$

$$= 2.50 \times 10^{-3} \text{ mol } H^+$$

The new concentration of $HC_2H_3O_2$ and $C_2H_3O_2^-$ are;

$$[HC_2H_3O_2]_{final} = \frac{(0.100 \text{ mol} - 0.00250 \text{ mol})}{0.525 \text{ L}} = 0.195 \text{ M}$$

$$[C_2H_3O_2^-]_{final} = \frac{(0.110 \text{ mol} + 0.00250 \text{ mol})}{0.525 \text{ L}} = 0.205 \text{ M}$$

Note: The new volume has been used in these calculations.

$$K_a = \frac{[H^+][C_2H_3O_2^-]}{[HC_2H_3O_2]} = \frac{[H^+](0.205)}{(0.195)} = 1.8 \times 10^{-5}$$

$[H^+] = 1.71 \times 10^{-5}$ and pH = 4.766.

Notice that the change in pH is very small in spite of adding a strong acid. If the same amount of HCl were added to water, a completely different effect would be observed.

Since HCl is a strong, the $[H^+]$ in a water solution will be the result of the strong acid dissociation. We do, of course, need to account for the dilution. Using the dilution equation, $M_1V_1 = M_2V_2$, we determine the $[H^+] = 4.76 \times 10^{-3}$ mol L^{-1} and the pH = 2.32. The change in pH in this case is $7.00 - 2.32 = 4.38$ pH units. A significantly larger change!!

16.84 Since HCO_2H and NaOH react in a 1:1 ratio:

$$HCO_2H + NaOH \rightarrow NaHCO_2 + H_2O$$

we can use the equation $V_a \times M_a = V_b \times M_b$ to determine the volume of NaOH that is required to reach the equivalence point, i.e. the point at which the number of moles of NaOH is equal to the number of moles of HCO_2H:

V_{NaOH} = 50 mL × 0.10/0.10 = 50 mL

Thus the final volume at the equivalence point will be 50 + 50 = 100 mL.

The concentration of $NaHCO_2$ would then be:

0.10 mol/L × 0.050 L = 5.0 X 10^{-3} mol HCO_2H = 5.0 X 10^{-3} mol $NaHCO_2$

5.0 X 10^{-3} mol/0.100 L = 5.0 X 10^{-2} M $NaHCO_2$

The hydrolysis of this salt at the equivalence point proceeds according to the following equilibrium: $HCO_2^- + H_2O \rightleftharpoons HCO_2H + OH^-$

$$K_b = \frac{[HCO_2H][OH^-]}{[HCO_2^-]} = 5.6 \times 10^{-11}$$

	$[HCO_2^-]$	$[HCO_2H]$	$[OH^-]$
I	0.050	–	–
C	–x	+x	+x
E	0.050–x	+x	+x

Substituting the above values for equilibrium concentrations into the mass action expression and assuming x << 0.050 gives:

$$K_b = \frac{(x)(x)}{0.050} = 5.6 \times 10^{-11}$$

$x^2 = 2.8 \times 10^{-12}$ ∴ x = 1.7 X 10^{-6} M = $[OH^-]$ = $[HCO_2H]$

pOH = –log$[OH^-]$ = –log(1.7 X 10^{-6}) = 5.77

pH = 14.00 – pOH = 14.00 – 5.77 = 8.23

Cresol red would be a good indicator, since it has a color change near the pH at the equivalence point.

16.87 a) $HC_2H_3O_2 \rightleftharpoons H^+ + C_2H_3O_2^-$ $K_a = \dfrac{[H^+][C_2H_3O_2^-]}{[HC_2H_3O_2]} = 1.8 \times 10^{-5}$

	$[HC_2H_3O_2]$	$[H^+]$	$[C_2H_3O_2^-]$
I	0.1000	–	–
C	–x	+x	+x
E	0.1000–x	+x	+x

Substituting the above values for equilibrium concentrations into the mass action expression and assuming that x << 0.1000 gives:

$$x = [H^+] = 1.342 \times 10^{-3} \text{ M}.$$
$$pH = -\log [H^+] = -\log (1.342 \times 10^{-3}) = 2.8724.$$

(b) When NaOH is added, it will react with the acetic acid present decreasing the amount in solution and producing additional acetate ion. Since this a one–to–one reaction, the number of moles of acetic acid will decrease by the same amount as the number of moles of NaOH added and the number of moles of acetate ion will increase by an identical amount. We must determine the number of moles of all ions present and calculate new concentrations accounting for dilution.

$$\text{\# moles } HC_2H_3O_2 = (0.02500 \text{ L solution})\left(\frac{0.1000 \text{ moles } HC_2H_3O_2}{1 \text{ L solution}}\right)$$
$$= 2.500 \times 10^{-3} \text{ moles } HC_2H_3O_2$$

$$\text{\# moles } OH^- = (0.01000 \text{ L solution})\left(\frac{0.1000 \text{ moles } OH^-}{1 \text{ L solution}}\right)$$
$$= 1.000 \times 10^{-3} \text{ moles } OH^-$$

$$[HC_2H_3O_2] = \frac{2.500 \times 10^{-3} \text{ moles} - 1.000 \times 10^{-3} \text{ moles}}{0.02500 \text{ L} + 0.01000 \text{ L}}$$
$$= 4.286 \times 10^{-2} \text{ M } HC_2H_3O_2$$

$$[C_2H_3O_2^-] = \frac{0 \text{ moles} + 1.000 \times 10^{-3} \text{ moles}}{0.02500 \text{ L} + 0.01000 \text{ L}}$$
$$= 2.857 \times 10^{-2} \text{ M } C_2H_3O_2^-$$

$$pH = pK_a + \log \frac{\left[C_2H_3O_2^-\right]}{\left[HC_2H_3O_2\right]}$$

$$= 4.7447 + \log \frac{\left(2.857 \times 10^{-2}\right)}{\left(4.286 \times 10^{-2}\right)} = 4.5686$$

(c)   When half the acetic acid has been neutralized, there will be equal amounts of acetic acid and acetate ion present in the solution. At this point, $pH = pK_a = 4.7447$.

(d)   At the equivalence point, all of the acetic acid will have been converted to acetate ion. The concentration of the acetate ion will be half the original concentration of acetic acid since we have doubled the volume of the solution. We then need to solve the equilibrium problem that results when we have a solution that possesses a $[C_2H_3O_2^-] = 0.05000$ M.

$$C_2H_3O_2^- + H_2O \rightleftharpoons HC_2H_3O_2 + OH^-$$

$$K_b = \frac{\left[HC_2H_3O_2\right]\left[OH^-\right]}{\left[C_2H_3O_2^-\right]} = 5.6 \times 10^{-10}$$

	$[C_2H_3O_2^-]$	$[HC_2H_3O_2]$	$[OH^-]$
I	0.05000	–	–
C	–x	+x	+x
E	0.05000–x	+x	+x

Substituting the above values for equilibrium concentrations into the mass action expression and assuming that $x \ll 0.05000$ gives: $x = [OH^-] = 5.292 \times 10^{-6}$ M.
$pOH = -\log [OH^-] = -\log (5.292 \times 10^{-6}) = 5.2764$.
$pH = 14.0000 - pOH = 14.0000 - 5.2764 = 8.7236$.

16.90   From the complete ionization of HCl: $[H^+] = 0.100$ M

We must solve an equilibrium problem in order to determine the contribution of H^+ from the acetic acid.

$$HC_2H_3O_2 \rightleftharpoons H^+ + C_2H_3O_2^- \qquad K_a = \frac{\left[H^+\right]\left[C_2H_3O_2^-\right]}{\left[HC_2H_3O_2\right]} = 1.8 \times 10^{-5}$$

	$[HC_2H_3O_2]$	$[H^+]$	$[C_2H_3O_2^-]$
I	0.125	0.100	–
C	–x	+x	+x
E	0.125–x	0.100+x	+x

Substituting the above values for equilibrium concentrations into the mass action expression and assuming that x << 0.125 M and x << 0.100 M gives:

$x = [C_2H_3O_2^-] = 2.25 \times 10^{-5}$ M.

The total $[H^+]$ is therefore $0.100 + 2.25 \times 10^{-5} = 0.100$ M, and pH is 1.000. The weak acid contributes very little H^+ to the final solution.

16.92 Only the cation is hydrolyzed:

$$NH_4^+ + H_2O \rightleftharpoons H_3O^+ + NH_3 \qquad K_a = \frac{[H_3O^+][NH_3]}{[NH_4^+]} = 5.7 \times 10^{-10}$$

	$[NH_4^+]$	$[H_3O^+]$	$[NH_3]$
I	0.120	–	–
C	–x	+x	+x
E	0.120–x	+x	+x

$$K_a = \frac{(x)(x)}{(0.120 - x)} = 5.7 \times 10^{-10}$$

Assuming x to be smaller than 0.120, we have: $x = 8.3 \times 10^{-6}$ M

$pH = -\log(8.3 \times 10^{-6}) = 5.08$

16.94 (a) The equation that will be used is the normal Henderson–Hasselbach equation, namely:

$$pH = pK_a + \log \frac{[anion]}{[acid]} = pK_a + \log \frac{[A^-]}{[HA]}$$

where $A^- = C_2H_3O_2^-$ and $HA = HC_2H_3O_2$. We note further that the log term involves a ratio of concentrations, but that the volume remains constant in a process such as that to be analyzed here. Thus the log term may be replaced by a ratio of mole amounts, since volumes cancel:

$$pH = pK_a + \log \frac{\left(\text{moles } C_2H_3O_2^-\right)}{\left(\text{moles } HC_2H_3O_2\right)}$$

Thus we need only determine the number of moles of acid and conjugate base that remain in the buffer after the addition of a certain amount of H^+ or OH^-, in order to determine the pH of the buffer mixture after that addition.

The buffer is changed in the following way by the addition of OH^-:

$$HC_2H_3O_2 + OH^- \rightleftharpoons H_2O + C_2H_3O_2^-$$

In other words, if 0.0100 moles of OH^- are added to the buffer, the amount of $HC_2H_3O_2$ goes down by 0.0100 moles, whereas the amount of $C_2H_3O_2^-$ goes up by 0.0100 moles. In addition, the pH of the solution will increase upon the addition of base.

The buffer is changed in the following way by the addition of H^+:

$$C_2H_3O_2^- + H^+ \rightleftharpoons HC_2H_3O_2$$

If 0.0100 mol of H^+ are added, then the amount of $C_2H_3O_2^-$ goes down by 0.0100 moles and the amount of $HC_2H_3O_2$ goes up by 0.0100 moles. As before, the addition of acid will decrease the pH of the buffer system.

For the general buffer mixture that contains x mol of $C_2H_3O_2^-$ and y mol of $HC_2H_3O_2$, we can apply the Henderson–Hasselbach equation, noting the maximum amount by which we want the pH to change (namely 0.10 units):

For the case of added base:

$$5.12 + 0.10 = 4.74 + \log \frac{(x + 0.0100) \text{ moles } C_2H_3O_2^-}{(x - 0.0100) \text{ moles } HC_2H_3O_2}$$

Simplifying and taking the antilog of both sides of the above equation gives: [x + 0.0100]/[y − 0.0100] = 3.0, which is now designated eq. 1.

For the case of added acid:

$$5.12 - 0.10 = 4.74 + \log \frac{(x - 0.0100) \text{ moles } C_2H_3O_2^-}{(x + 0.0100) \text{ moles } HC_2H_3O_2}$$

Simplifying and taking the antilog of both sides of the equation gives:
[x – 0.0100]/[y + 0.0100] = 1.9, which is now designated eq. 2.

The equations 1 and 2 as designated above are solved simultaneously, since they are two equations containing two unknowns:

x = initial mol $C_2H_3O_2^-$ = 0.15 mol

y = initial mol $HC_2H_3O_2$ = 6.3 X 10^{-2} mol

These values are converted to grams as follows:

6.3 X 10^{-2} mol × 60.1 g/mol = 3.8 g $HC_2H_3O_2$

0.15 mol $NaC_2H_3O_2$ × 118 g/mol = 18 g $NaC_2H_3O_2 \cdot 2H_2O$

These are the minimum amounts of acid and conjugate base that would be required in order to prepare a buffer that would change pH by only 0.10 units on addition of either 0.0100 mol of OH^- or 0.0100 mol of H^+.

(b)　6.3 X 10^{-2} mol/0.250 L = 0.25 M $HC_2H_3O_2$

　　0.15 mol $C_2H_3O_2^-$/0.250 L = 0.60 M $NaC_2H_3O_2$

(c)　0.0100 mol OH^-/0.250 L = 4.00 X 10^{-2} M OH^-

　　pOH = –log[OH^-] = –log(4.00 X 10^{-2}) = 1.398

　　pH = 14.000 – pOH = 14.000 – 1.398 = 12.602

(d)　1.00 X 10^{-2} mol H^+/0.250 L = 4.00 X 10^{-2} M H^+

　　pH = –log[H^+] = –log(4.0 X 10^{-2}) = 1.398

16.95　Since the ammonium ion is the salt of a weak base, NH_3, it is acidic. The cyanide ion is the salt of a weak acid, HCN, so it is basic. In order to determine if the solution is acidic or basic, we need to determine the relative strength of the two components. Use the relationship $K_a \times K_b = K_w$ in order to determine the dissociation constants for the cyanide ion and the ammonium ion.

$$K_a(NH_4^+) = K_w \div K_b(NH_3) = 1.0 \text{ X } 10^{-14} \div 1.8 \text{ X } 10^{-5} = 5.6 \text{ X } 10^{-10}$$

$$K_b(CN^-) = K_w \div K_a(HCN) = 1.0 \text{ X } 10^{-14} \div 6.2 \text{ X } 10^{-9} = 1.6 \text{ X } 10^{-5}$$

Since the $K_b(CN^-)$ is larger than the $K_a(NH_4^+)$ the NH_4CN solution will be basic.

16.97 In order to solve this problem we must first neutralize the HNO_3 that is present in the solution. Since HNO_3 is a monoprotic acid and NH_3 is a monobasic base, they will react in a one to one stoichiometry. For every mole of NH_3 added, one mole of HNO_3 will be neutralized and one mole of NH_4^+ will be produced. There are initially 0.0125 moles $HNO_3 = (0.250 \text{ L})(0.050 \text{ mol L}^{-1})$ present in the solution. Consequently, we must first add 0.0125 moles of NH_3 to neutralize all of the HNO_3.

Now the question may be rephrased to ask, how much ammonia should be added to a solution that is 0.050 M in NH_4^+ so that the final pH is 9.26?

	$[NH_3]$	$[NH_4^+]$	$[OH^-]$
I	Z	0.050	–
C	Z–x	0.050+x	+x
E	Z–x	0.050+x	+x

$$K_b = \frac{(0.050+x)(x)}{(Z-x)} = 1.8 \times 10^{-5}$$

The problem states that the equilibrium pH = 9.26 and, therefore, the pOH = 4.74 which implies that $[OH^-] = 1.8 \times 10^{-5} \text{ M} = x$. Substituting this into the equilibrium expression and assuming that $Z \gg x$ we get

$$\frac{(0.050)(1.8 \times 10^{-5})}{Z} = 1.8 \times 10^{-5}$$

and Z = 0.050 M. Since the volume of the solution is 250 mL, this concentration corresponds to 0.0125 moles $NH_3 = (0.250 \text{ L})(0.050 \text{ mol L}^{-1})$.

The total amount of NH_3 that must be added to the original solution is thus 0.0125 moles + 0.0125 moles = 0.0250 mols. The first addition neutralizes the HNO_3 and the second amount adjusts the pH.

The volume of NH_3 that must be added may be calculated using the ideal gas law;

$$V = \frac{nRT}{P} = \frac{(0.0250 \text{ mol})\left(0.0821 \frac{L \text{ atm}}{mol \text{ K}}\right)(298 \text{ K})}{(740 \text{ torr})\left(\frac{1 \text{ atm}}{760 \text{ torr}}\right)} = 0.63 \text{ L} = 630 \text{ mL}$$

CHAPTER SEVENTEEN
Review Exercises

17.1 (a) $H_3AsO_4 \rightleftharpoons H_2AsO_4^- + H^+$

$H_2AsO_4^- \rightleftharpoons HAsO_4^{2-} + H^+$

$HAsO_4^{2-} \rightleftharpoons AsO_4^{3-} + H^+$

(b) $K_1 = 10^{-2.2} = 6 \times 10^{-3}$

$K_2 = 10^{-6.9} = 1 \times 10^{-7}$

$K_3 = 10^{-11.5} = 3 \times 10^{-12}$

(c) $HAsO_4^{2-}$ is the stronger acid because it has the larger K_a (See Table 17.1).

17.3 (a) $H_2CO_3 \rightleftharpoons HCO_3^- + H^+$

$$K_{a_1} = \frac{\left[HCO_3^-\right]\left[H^+\right]}{\left[H_2CO_3\right]}$$

$HCO_3^- \rightleftharpoons CO_3^{2-} + H^+$

$$K_{a_2} = \frac{\left[CO_3^{2-}\right]\left[H^+\right]}{\left[HCO_3^-\right]}$$

(b) According to Table 17.1, sulfurous acid has a larger acid ionization constant and is, therefore, the stronger acid.

17.5 $H_3PO_3 \rightleftharpoons H_2PO_3^- + H^+$

$$K_{a_1} = \frac{\left[H_2PO_3^-\right]\left[H^+\right]}{\left[H_3PO_3\right]} = 1.0 \times 10^{-2}$$

$H_2PO_3^- \rightleftharpoons HPO_3^{2-} + H^+$

$$K_{a_2} = \frac{\left[HPO_3^{2-}\right]\left[H^+\right]}{\left[H_2PO_3^-\right]} = 2.6 \times 10^{-7}$$

To simplify the calculation, assume that the second dissociation does not contribute a significant amount of H^+ to the final solution. Solving the equilibrium problem for the first dissociation gives:

	$[H_3PO_3]$	$[H_2PO_3^-]$	$[H^+]$
I	1.0	–	–
C	–x	+x	+x
E	1.0–x	+x	+x

$$K_{a_1} = \frac{[H_2PO_3^-][H^+]}{[H_3PO_3]} = \frac{(x)(x)}{(1.0-x)} = 1.0 \times 10^{-2}$$

Because K_{a_1} is so large, a quadratic equation must be solved. On doing so we learn that $x = 0.095\ M = [H^+] = [H_2PO_3^-]$.

$pH = -\log[H^+] = -\log(0.095) = 1.022$

The $[HPO_3^{2-}]$ may be determined from the second ionization constant.

	$[H_2PO_3^-]$	$[HPO_3^{2-}]$	$[H^+]$
I	0.095	–	0.095
C	–x	+x	+x
E	0.095–x	+x	0.095+x

$$K_{a_2} = \frac{[HPO_3^{2-}][H^+]}{[H_2PO_3^-]} = \frac{(x)(0.95+x)}{(0.095-x)} = 2.6 \times 10^{-7}$$

If we assume that x is small then $0.095 \pm x \approx 0.095$. Then, $x = [HPO_3^{2-}] = 2.6 \times 10^{-7}\ M$.

17.7 The equilibrium is:

$$H_2Tar \rightleftharpoons HTar^- + H^+ \qquad K_{a_1} = \frac{[HTar^-][H^+]}{[H_2Tar]} = 9.2 \times 10^{-4}$$

$$HTar^- \rightleftharpoons Tar^{2-} + H^+ \qquad K_{a_2} = \frac{[Tar^{2-}][H^+]}{[HTar^-]} = 4.3 \times 10^{-5}$$

The $[H^+]$ and $[HTar^-]$ are determined by the first dissociation equation.

	$[H_2Tar]$	$[HTar^-]$	$[H^+]$
I	0.10	–	–
C	–x	+x	+x
E	0.10–x	+x	+x

$$K_{a_1} = \frac{[HTar^-][H^+]}{[H_2Tar]} = \frac{(x)(x)}{(0.10 - x)} = 9.2 \times 10^{-4}$$

Solving this equation requires a quadratic equation or the method of successive approximations:

$$x = 0.0091\ M = [H^+] = [HTar^-]$$
$$[Tar^{2-}] = K_{a_2} = 4.3 \times 10^{-5}\ M$$

17.9 The dissociation that produces the H^+ also produces an equivalent amount of HCO_3^{2-}. Consequently, the $[HCO_3^{2-}] = [H^+] = 10^{-pH} = 3 \times 10^{-4}\ M$.

$$[CO_3^{2-}] = K_{a_2} = 4.7 \times 10^{-11}\ M.$$

17.11 $[C_4H_4O_6^{2-}] = K_{a_2} = 4.3 \times 10^{-5}\ M.$

17.13 (a) $(NH_4)_2SO_4$, $NaHSO_4$
(b) Na_3PO_4, $LiHCO_3$, K_2HPO_4, K_2SO_3
(c) none

17.15 The hydrolysis equation is:

$$SO_3^{2-} + H_2O \rightleftharpoons HSO_3^- + OH^- \qquad K_b = \frac{\left[HSO_3^-\right]\left[OH^-\right]}{\left[SO_3^{2-}\right]}$$

In order to obtain K_b we will use the relationship $K_w = K_a \times K_b$

$$K_b = \frac{K_w}{K_a} = \frac{1.0 \times 10^{-14}}{6.6 \times 10^{-8}} = 1.5 \times 10^{-7}$$

	$[SO_3^{2-}]$	$[HSO_3^-]$	$[OH^-]$
I	0.12	–	–
C	–x	+x	+x
E	0.12–x	+x	+x

Since K_b is so small, assume that $x \ll 0.12$ and we determine $x = 1.3 \times 10^{-4}$ M $= [OH^-]$

$pOH = -\log(1.3 \times 10^{-4}) = 3.87$, $pH = 14.00 - pOH = 10.13$

17.17 Like problem number 16, only the first hydrolysis needs to be examined. The difficulty is that we are solving for the initial concentration in this problem.

$$CO_3^{2-} + H_2O \rightleftharpoons HCO_3^- + OH^- \qquad K_b = \frac{\left[HCO_3^-\right]\left[OH^-\right]}{\left[CO_3^{2-}\right]}$$

In order to obtain K_b we will use the relationship $K_w = K_a \times K_b$

$$K_b = \frac{K_w}{K_a} = \frac{1.0 \times 10^{-14}}{4.7 \times 10^{-11}} = 2.1 \times 10^{-4}$$

	$[CO_3^{2-}]$	$[HCO_3^-]$	$[OH^-]$
I	Z	–	–
C	–x	+x	+x
E	Z–x	+x	+x

In this problem, we know that the pH at equilibrium is 11.62. Using this fact we calculate a $pOH = 2.38$ and a $[OH^-] = 10^{-pOH} = 4.2 \times 10^{-3}$ M $= x$. Substitute this into the equilibrium expression and solve for **Z**, our initial concentration of carbonate ion.

$$K_b = 2.1 \times 10^{-4} = \frac{\left[HCO_3^-\right]\left[OH^-\right]}{\left[CO_3^{2-}\right]} = \frac{(x)(x)}{(Z-x)} = \frac{\left(4.2 \times 10^{-3}\right)^2}{Z - 4.2 \times 10^{-3}}$$

Solving this equation we determine that $Z = 8.8 \times 10^{-2}$ M $= [CO_3^{2-}]$. (Note: We could have assumed that the change upon dissociation is small in this example, i.e., $x \ll Z$. It is unnecessary to do this for this problem. If you had made this assumption, a value of $Z = 8.4 \times 10^{-2}$ M would be obtained.)

The problem asks for the number of grams of $Na_2CO_3 \bullet 10H_2O$ so;

$$\# \text{ g } Na_2CO_3 \bullet 10H_2O = (1\,L)\left(\frac{8.8 \times 10^{-2} \text{ moles } CO_3^{2-}}{L}\right)\left(\frac{1 \text{ mole } Na_2CO_3 \bullet 10H_2O}{1 \text{ mole } CO_3^{2-}}\right)$$

$$\times \left(\frac{286 \text{ g } Na_2CO_3 \bullet 10H_2O}{1 \text{ mole } Na_2CO_3 \bullet 10H_2O}\right)$$

$$= 25 \text{ g } Na_2CO_3 \bullet 10H_2O$$

17.19   Because the pH is so high and is greater than pK_a, the $\left[CO_3^{2-}\right] = 0.10$ M.

17.22   (a) $CaF_2(s) \rightleftharpoons Ca^{2+} + 2F^-$ $\qquad K_{sp} = \left[Ca^{2+}\right]\left[F^-\right]^2$

(b) $Ag_2CO_3(s) \rightleftharpoons 2Ag^+ + CO_3^{2-}$ $\qquad K_{sp} = \left[Ag^+\right]^2\left[CO_3^{2-}\right]$

(c) $PbSO_4(s) \rightleftharpoons Pb^{2+} + SO_4^{2-}$ $\qquad K_{sp} = \left[Pb^{2+}\right]\left[SO_4^{2-}\right]$

(d) $Fe(OH)_3(s) \rightleftharpoons Fe^{3+} + 3OH^-$ $\qquad K_{sp} = \left[Fe^{3+}\right]\left[OH^-\right]^3$

(e) $PbI_2(s) \rightleftharpoons Pb^{2+} + 2I^-$ $\qquad K_{sp} = \left[Pb^{2+}\right]\left[I^-\right]^2$

(f) $Cu(OH)_2(s) \rightleftharpoons Cu^{2+} + 2OH^-$ $\qquad K_{sp} = \left[Cu^{2+}\right]\left[OH^-\right]^2$

17.26 (a)

$$\text{\# moles BaSO}_4 = \left(0.00245 \text{ g BaSO}_4\right)\left(\frac{1 \text{ mole BaSO}_4}{233.3906 \text{ g BaSO}_4}\right)$$

$$\text{\# moles BaSO}_4 = 1.05 \text{ X } 10^{-5} \text{ moles}$$

(b) $\left[Ba^{2+}\right] = \left[SO_4^{2-}\right] = 1.05 \text{ X } 10^{-5} \text{ M}$

(c) $K_{sp} = \left[Ba^{2+}\right]\left[SO_4^{2-}\right] = (1.05 \text{ X } 10^{-5})^2 = 1.10 \text{ X } 10^{-10}$

17.28

$$\text{\# moles AgC}_2\text{H}_3\text{O}_2 = \left(0.800 \text{ g AgC}_2\text{H}_3\text{O}_2\right)\left(\frac{1 \text{ mole AgC}_2\text{H}_3\text{O}_2}{166.9 \text{ g AgC}_2\text{H}_3\text{O}_2}\right)$$

$$= 4.79 \text{ X } 10^{-3} \text{ moles AgC}_2\text{H}_3\text{O}_2$$

$[AgC_2H_3O_2] = 4.79 \text{ X } 10^{-3} \text{ moles AgC}_2\text{H}_3\text{O}_2 / 0.100 \text{ L} = 4.79 \text{ X } 10^{-2} \text{ M}$

One mole of Ag^+ and one mole of $C_2H_3O_2$ will be produced for every mole of

$AgC_2H_3O_2$. Therefore; $K_{sp} = \left[Ag^+\right]\left[C_2H_3O_2^-\right] = \left(4.79 \text{ X } 10^{-2}\right)^2 = 2.29 \text{ X } 10^{-3}$.

17.30 $BaSO_3 \text{ (s)} \rightleftharpoons Ba^{2+} + SO_3^{2-}$ $\qquad$ $K_{sp} = \left[Ba^{2+}\right]\left[SO_3^{2-}\right]$

$K_{sp} = (0.10)(8.0 \text{ X } 10^{-6}) = 8.0 \text{ X } 10^{-7}$

In this problem, all of the Ba^{2+} comes from the $BaCl_2$.

17.32 (a) $CuCl(s) \rightleftharpoons Cu^+(aq) + Cl^-(aq)$ $\qquad$ $K_{sp} = \left[Cu^+\right]\left[Cl^-\right]$

	$[Cu^+]$	$[Cl^-]$
I	–	–
C	+x	+x
E	+x	+x

$K_{sp} = x^2 = 1.9 \text{ X } 10^{-7}$ $\therefore$ x = molar solubility = $4.4 \text{ X } 10^{-4}$

(b) $CuCl(s) \rightleftharpoons Cu^+(aq) + Cl^-(aq)$ $K_{sp} = \left[Cu^+\right]\left[Cl^-\right]$

	$[Cu^+]$	$[Cl^-]$
I	–	0.0200
C	+x	+x
E	+x	0.0200+x

$K_{sp} = (x)(0.0200+x) = 1.9 \times 10^{-7}$ Assume that x << 0.0200

$\therefore$ x = molar solubility = 9.5×10^{-6} M

(c) $CuCl(s) \rightleftharpoons Cu^+(aq) + Cl^-(aq)$ $K_{sp} = \left[Cu^+\right]\left[Cl^-\right]$

	$[Cu^+]$	$[Cl^-]$
I	–	0.200
C	+x	+x
E	+x	0.200+x

$K_{sp} = (x)(0.200+x) = 1.9 \times 10^{-7}$ Assume that x << 0.200

$\therefore$ x = molar solubility = 9.5×10^{-7} M

(d) $CuCl(s) \rightleftharpoons Cu^+(aq) + Cl^-(aq)$ $K_{sp} = \left[Cu^+\right]\left[Cl^-\right]$

Note that the Cl^- concentration equals (2)(0.150 M) since two moles of Cl^- are produced for every mole of $CaCl_2$.

	$[Cu^+]$	$[Cl^-]$
I	–	0.300
C	+x	+x
E	+x	0.300+x

$K_{sp} = (x)(0.300+x) = 1.9 \times 10^{-7}$ Assume that x << 0.300

$\therefore$ x = molar solubility = 6.3×10^{-7} M

Chapter Seventeen

17.34 To solve this problem, determine the molar solubility for each compound.

LiF: let $x = \left[Li^+\right] = \left[F^-\right]$ $K_{sp} = \left[Li^+\right]\left[F^-\right] = x^2 = 1.7 \times 10^{-3}$

$x = 4.1 \times 10^{-2}$ moles/L = molar solubility of LiF.

BaF_2: let $x = \left[Ba^{2+}\right]$, $\left[F^-\right] = 2x$ $K_{sp} = \left[Ba^{2+}\right]\left[F^-\right]^2 = (x)(2x)^2 = 1.7 \times 10^{-6}$

$4x^3 = 1.7 \times 10^{-6}$, and $x = 7.5 \times 10^{-3}$ M = molar solubility of BaF_2.

Because the molar solubility of LiF is greater than the molar solubility of BaF_2, LiF is more soluble.

17.36 First determine the molar solubility of the MX salt.

Let $x = \left[M^+\right] = \left[X^-\right]$, $K_{sp} = \left[M^+\right]\left[X^-\right] = (x)(x) = 3.2 \times 10^{-10}$

$x = 1.8 \times 10^{-5}$ M. This is the equilibrium concentration of the two ions.

For the MX_3 salt, let x = equilibrium concentration of M^{3+}, $\left[X^-\right] = 3x$.

$K_{sp} = \left[M^{3+}\right]\left[X^-\right]^3 = (x)(3x)^3 = 27x^4$. The value of x in this expression is the value determined in the first part of this problem.

So, $K_{sp} = (27)(1.8 \times 10^{-5})^4 = 2.8 \times 10^{-18}$

17.38 $CaSO_4(s) \rightleftharpoons Ca^{2+}(aq) + SO_4^{2-}(aq)$ $K_{sp} = \left[Ca^{2+}\right]\left[SO_4^{2-}\right]$

let $x = \left[Ca^{2+}\right] = \left[SO_4^{2-}\right]$ $K_{sp} = x^2 = 2.4 \times 10^{-5}$ and $x = 4.9 \times 10^{-3}$ M.

The molar solubility of $CaSO_4$ is 4.9×10^{-3} moles/L.

17.40 For every mole of Pb^{2+} produced, 2 moles of I^- will be produced. Let $x = \left[Pb^{2+}\right]$ at equilibrium and $\left[I^-\right] = 2x$ at equilibrium. $K_{sp} = \left[Pb^{2+}\right]\left[I^-\right]^2 = (x)(2x)^2 = 4x^3$. Solving we find $x = 1.3 \times 10^{-3}$ M. Thus, the molar solubility of PbI_2 is 1.3×10^{-3} moles/L.

17.42 $Ag_2CrO_4 (s) \rightleftharpoons 2Ag^+ (aq) + CrO_4^{2-}(aq)$ $K_{sp} = \left[Ag^+\right]^2\left[CrO_4^{2-}\right]$

	$[Ag^+]$	$[CrO_4^{2-}]$
I	0.200	–
C	0.200+2x	+x
E	0.200+2x	+x

$K_{sp} = (0.200+2x)^2(x)$ Assume that x << 0.200

$1.2 \times 10^{-12} = (0.200)^2(x)$ $x = 3.0 \times 10^{-11}$

The molar solubility is 3.0×10^{-11} moles/L.

17.44 $MgOH_2 (s) \rightleftharpoons Mg^{2+} (aq) + 2OH^-(aq)$ $K_{sp} = \left[Mg^{2+}\right]\left[OH^-\right]^2$

	$[Mg^{2+}]$	$[OH^-]$
I	–	0.20
C	+x	0.20+2x
E	+x	0.20+2x

$K_{sp} = (x)(0.20+2x)^2$ Assume 2x << 0.20

$K_{sp} = (x)(0.20)^2 = 7.1 \times 10^{-12}$ $x = 1.8 \times 10^{-10}$ M

The assumption is okay and the molar solubility is 1.8×10^{-10} moles/L

17.46 In order for a precipitate to form, the value of the reaction quotient, Q, must be greater than the value of K_{sp}. For $AgC_2H_3O_2$, $K_{sp} = 2.3 \times 10^{-3}$.

$Q = \left[Ag^+\right]\left[C_2H_3O_2^-\right] = (0.015)(0.50) = 7.5 \times 10^{-3}$. Since $Q > K_{sp}$, a precipitate will form. (Note: The concentration of $C_2H_3O_2^-$ is twice the concentration of $Ca(C_2H_3O_2)_2$ since one mole of $Ca(C_2H_3O_2)_2$ produces two moles of $C_2H_3O_2^-$).

17.48   In order for a precipitate to form, the value of the reaction quotient, Q, must be greater than the value of K_{sp}. For $AgC_2H_3O_2$, $K_{sp} = 2.3 \times 10^{-3}$.

$$\left[Ag^+\right] = (22.0 \text{ mL})(0.100 \text{ M})/(67.0 \text{ mL}) = 3.28 \times 10^{-2} \text{ M}$$

$$\left[C_2H_3O_2^-\right] = (45.0 \text{ mL})(0.0260 \text{ M})/(67.0 \text{ mL}) = 1.75 \times 10^{-2} \text{ M}$$

$$Q = \left[Ag^+\right]\left[C_2H_3O_2^-\right] = (3.28 \times 10^{-2})(1.75 \times 10^{-2}) = 5.73 \times 10^{-4}.$$

Since $Q < K_{sp}$, no precipitate will form.

17.50 $CaSO_4$ (s) $\rightleftharpoons$ Ca^{2+} (aq) + SO_4^{2-} (aq) $K_{sp} = \left[Ca^{2+}\right]\left[SO_4^{2-}\right]$

let $x = [Ca^{2+}]$, $[SO_4^{2-}] = 0.015 \text{ M} + x$ $K_{sp} = (x)(0.015 + x) = 2.4 \times 10^{-5}$

Solving the resulting quadratic equation gives $x = 1.5 \times 10^{-3}$ M.

The molar solubility of $CaSO_4$ in 0.015 M $CaCl_2$ is 1.5×10^{-3} moles/L.

(Note: If we had assumed that $x \ll 0.015$, as is usually the method we use to solve these problems, we would have determined that $x = 0.0016$ M. This is slightly larger than 10% of the value 0.015 so our usual assumption is not valid in this problem.)

17.55   The less soluble substance is PbS. We need to determine the minimum $[H^+]$ at which CoS will precipitate.

$$K_{spa} = \frac{\left[Co^{2+}\right]\left[H_2S\right]}{\left[H^+\right]^2} = \frac{(0.010)(0.1)}{[H^+]^2} = 0.5 \text{ (from Table 17.3)}$$

$$[H^+] = \sqrt{\frac{(0.010)(0.1)}{0.5}} = 0.045$$

$pH = -\log[H^+] = 1.35$. At a pH lower than 1.35, PbS will precipitate and CoS will not. At larger values of pH, both PbS and CoS will precipitate.

17.57   (a) $\# \text{ mol}/L \left(7.05 \times 10^{-3} \text{ g}/L\right)\left(\dfrac{1 \text{ mol Mg(OH)}_2}{58.33 \text{ g Mg(OH)}_2}\right) = 1.21 \times 10^{-4} \text{ M}$

(b) $\left[Mg^{2+}\right] = 1.21 \times 10^{-4} \text{ M}$ $\left[OH^-\right] = 2.42 \times 10^{-4} \text{ M}$

(c) $K_{sp} = \left[Mg^{2+}\right]\left[OH^-\right]^2 = \left(1.21 \times 10^{-4}\right)\left(2.42 \times 10^{-4}\right)^2 = 7.09 \times 10^{-12}$

17.58 (a)   The number of moles of the two reactants are:

$$0.12 \text{ M Ag}^+ \times 0.050 \text{ L} = 6.0 \times 10^{-3} \text{ moles Ag}^+$$
$$0.048 \text{ M Cl}^- \times 0.050 \text{ L} = 2.4 \times 10^{-3} \text{ moles Cl}^-$$

The precipitation of AgCl proceeds according to the following stoichiometry: $Ag^+ + Cl^- \rightarrow AgCl(s)$. If we assume that the product is completely insoluble, then 2.4×10^{-3} moles of AgCl will be formed because Cl^- is the limiting reagent (see above.)

$$\# \text{ g AgCl} = \left(2.4 \times 10^{-3} \text{ mol AgCl}\right)\left(\frac{143.3 \text{ g AgCl}}{1 \text{ mol AgCl}}\right) = 0.35 \text{ g AgCl}$$

(b)   The silver ion concentration may be determined by calculating the amount of excess silver added to the solution:

$$[Ag^+] = (6.0 \times 10^{-3} \text{ moles} - 3.6 \times 10^{-2} \text{ moles})/1.00 \text{ L} = 3.6 \times 10^{-2} \text{ M}$$

The concentrations of nitrate and sodium ions are easily calculated since they are spectators in this reaction:

$$[NO_3^-] = (0.12 \text{ M})(50.0 \text{ mL})/(100.0 \text{ mL}) = 6.0 \times 10^{-2} \text{ M}$$
$$[Na^+] = (0.048 \text{ M})(50.0 \text{ mL})/(100.0 \text{ mL}) = 2.4 \times 10^{-2} \text{ M}$$

In order to determine the chloride ion concentration, we need to solve the equilibrium expression. Specifically, we need to ask what is the chloride ion concentration in a saturated solution of AgCl that has a $[Ag^+] = 3.6 \times 10^{-2}$ M.

$$AgCl(s) \rightleftharpoons Ag^+ + Cl^- \qquad K_{sp} = 1.8 \times 10^{-10}$$

	AgCl	$[Ag^+]$	$[Cl^-]$
I		0.036	–
C		+x	+x
E		0.036+x	+x

$$K_{sp} = \left[Ag^+\right]\left[Cl^-\right] = (0.036+x)(x) = 1.8 \times 10^{-10}$$

$$x = 5.0 \times 10^{-9} \text{ M if we assume that } x<<0.036$$

Therefore, $[Cl^-] = 5.0 \times 10^{-9}$ M.

(c)   The percentage of the silver that has precipitated is:

$$(2.4 \times 10^{-3} \text{ moles})/(6.0 \times 10^{-3} \text{ moles}) \times 100\% = 40\%$$

17.60   A saturated solution of $La_2(CO_3)_3$ satisfies the following equilibrium expression:

$$K_{sp} = 4.0 \times 10^{-34} = \left[La^{3+}\right]^2\left[CO_3^{2-}\right]^3$$

If $\left[La^{3+}\right] = 0.010$ M, then the carbonate concentration of a saturated solution is:

$$\left[CO_3^{2-}\right] = \sqrt[3]{\frac{K_{sp}}{\left[La^{3+}\right]^2}} = \sqrt[3]{\frac{4.0 \times 10^{-34}}{(0.010)^2}} = 1.6 \times 10^{-10}$$

We do the same calculation for $PbCO_3$:

$$K_{sp} = 7.4 \times 10^{-14} = \left[Pb^{2+}\right]\left[CO_3^{2-}\right]$$

If $\left[Pb^{2+}\right] = 0.010$ M, then the carbonate concentration of a saturated solution is:

$$\left[CO_3^{2-}\right] = \frac{K_{sp}}{\left[Pb^{2+}\right]} = \frac{7.4 \times 10^{-14}}{0.010} = 7.4 \times 10^{-12} \text{ M}$$

Therefore, at a carbonate ion concentration between 7.4×10^{-12} M and 1.6×10^{-10} M, $PbCO_3$ will precipitate, but $La_2(CO_3)_3$ will not precipitate. The upper limit for the carbonate ion concentration is therefore 1.6×10^{-10} M.

The equilibrium we need to look at now is: $H_2CO_3(aq) \rightleftharpoons 2H^+(aq) + CO_3^{2-}(aq)$

The K_a for this reaction is the product of K_{a_1} and K_{a_2} for carbonic acid. From Table 17.1 we see that $K_{a_1} = 4.5 \times 10^{-7}$ and $K_{a_2} = 4.7 \times 10^{-11}$. So the equilibrum expression and value for the reaction of interest is:

$$K_a = \frac{\left[H^+\right]^2\left[CO_3^{2-}\right]}{\left[H_2CO_3\right]} = K_{a1} \times K_{a2} = 2.1 \times 10^{-17}$$

This equation is rearranged and the values above and the values given in the problem are substituted in order to determine the pH range over which $PbCO_3$ will selectively precipitate:

$$\left[H^+\right] = \sqrt{\frac{K_a\left[H_2CO_3\right]}{\left[CO_3^{2-}\right]}} = \sqrt{\frac{\left(2.1 \times 10^{-17}\right)\left(3.3 \times 10^{-2}\right)}{\left[CO_3^{2-}\right]}}$$

If we substitute $\left[CO_3^{2-}\right] = 7.4 \times 10^{-12}$ M we determine $\left[H^+\right] = 3.1 \times 10^{-4}$ M and the pH = 3.51. Substituting $\left[CO_3^{2-}\right] = 1.6 \times 10^{-10}$ M, $\left[H^+\right] = 6.6 \times 10^{-5}$ M and pH = 4.18.

Consequently, if $[H^+] = 6.6 \times 10^{-5}$ M (pH = 4.18), $La_2(CO_3)_3$ will not precipitate but $PbCO_3$ will precipitate. At pH = 3.51 and below, neither carbonate will precipitate.

18.13 (a)   Water is more readily oxidized than NO_3^-, so we have:

$$2H_2O(\ell) \rightarrow 4H^+(aq) + 4e^- + O_2(g)$$

   (b) Br^- is more readily oxidized than water, so we have:

$$2Br^-(aq) \rightarrow Br_2(aq) + 2e^-$$

   (c) $Br^-(aq)$ is more readily oxidized than NO_3^-, so we have:

$$2Br^-(aq) \rightarrow Br_2(aq) + 2e^-$$

18.15 The answers to the previous Review Exercises guide us here:
At the cathode, where reduction occurs, we expect $Cu(s)$. At the anode, where oxidation occurs, we expect $Br_2(aq)$.

18.17 $1\ C = 1\ A{\cdot}s$

   (a) $4.00\ A \times 600\ s = 2.40 \times 10^3\ C$

   (b) $10.0\ A \times 20\ min \times 60\ s/min = 1.2 \times 10^4\ C$

   (c) $1.50\ A \times 6.00\ hr \times 3600\ s/hr = 3.24 \times 10^4\ C$

18.19 (a) $Fe^{2+}(aq) + 2e^- \rightarrow Fe(s)$
          $0.20\ mol\ Fe^{2+} \times 2\ mol\ e^-/mol\ Fe^{2+} = 0.40\ mol\ e^-$

   (b) $Cl^-(aq) \rightarrow 1/2Cl_2(g) + e^-$

          $0.70\ mol\ Cl^- \times 1\ mol\ e^-/mol\ Cl^- = 0.70\ mol\ e^-$

   (c) $Cr^{3+}(aq) + 3e^- \rightarrow Cr(s)$
          $1.50\ mol\ Cr^{3+} \times 3\ mol\ e^-/mol\ Cr^{3+} = 4.50\ mol\ e^-$

   (d) $Mn^{2+}(aq) + 4H_2O(\ell) \rightarrow MnO_4^-(aq) + 8H^+(aq) + 5e^-$

          $1.0 \times 10^{-2}\ mol\ Mn^{2+} \times 5\ mol\ e^-/mol\ Mn^{2+} = 5.0 \times 10^{-2}\ mol\ e^-$

18.21 $Ag^+(aq) + e^- \rightarrow Ag(s)$, and $Cr^{3+}(aq) + 3e^- \rightarrow Cr(s)$
This shows that there are three moles of electrons per mole of Cr but only one mole of electrons per mole of Ag. The number of moles of electrons involved in the silver reaction is:

$$\#\ mol\ e^- = \left(12.0\ g\ Ag\right)\left(\frac{1\ mol\ Ag}{107.9\ g\ Ag}\right)\left(\frac{1\ mol\ e^-}{1\ mol\ Ag}\right) = 0.111\ mol\ e^-$$

The amount of Cr is then:

$$\# \text{ mol Cr}^{3+} = \left(0.111 \text{ mol e}^-\right)\left(\frac{1 \text{ mol Cr}^{3+}}{3 \text{ mol e}^-}\right) = 0.0371 \text{ mol Cr}^{3+}$$

18.23 $\text{Fe(s)} + 2\text{OH}^-(\text{aq}) \rightarrow \text{Fe(OH)}_2(\text{s}) + 2\text{e}^-$

The number of Coulombs is: $12.0 \text{ min} \times 60 \text{ s/min} \times 8.00 \text{ C/s} = 5.76 \text{ X } 10^3 \text{ C}$. The number of grams of Fe(OH)_2 is:

$$\# \text{ g Fe(OH)}_2 = \left(5.76 \text{ X } 10^3 \text{ C}\right)\left(\frac{1 \text{ mol e}^-}{96500 \text{ C}}\right)\left(\frac{1 \text{ mol Fe(OH)}_2}{2 \text{ mol e}^-}\right)\left(\frac{89.86 \text{ g Fe(OH)}_2}{1 \text{ mol Fe(OH)}_2}\right)$$

$$= 2.68 \text{ g Fe(OH)}_2$$

18.25   The electrolysis of NaCl solution results in the reduction of water, together with the formation of hydroxide ion: $2\text{H}_2\text{O}(\ell) + 2\text{e}^- \rightarrow \text{H}_2(\text{g}) + 2\text{OH}^-(\text{aq})$. The number of Coulombs is: $2.50 \text{ A} \times 15.0 \text{ min} \times 60 \text{ s/min} = 2.25 \text{ X } 10^3 \text{ C}$. The number of moles of OH^- is:

$$\# \text{ mol OH}^- = \left(2.25 \text{ X } 10^3 \text{ C}\right)\left(\frac{1 \text{ mol e}^-}{96500 \text{ C}}\right)\left(\frac{2 \text{ mol OH}^-}{2 \text{ mol e}^-}\right) = 0.0233 \text{ mol OH}^-$$

The volume of acid solution that will neutralize this much OH^- is:

$$\# \text{ mL HCl} = \left(0.0233 \text{ mol OH}^-\right)\left(\frac{1 \text{ mol HCl}}{1 \text{ mol OH}^-}\right)\left(\frac{1000 \text{ mL HCl}}{0.100 \text{ mol HCl}}\right) = 233 \text{ mL HCl}$$

18.27   The number of Coulombs that will be required is:

$$\# \text{ C} = \left(35.0 \text{ g Pb}\right)\left(\frac{1 \text{ mol Pb}}{207.2 \text{ g Pb}}\right)\left(\frac{2 \text{ mol e}^-}{1 \text{ mol Pb}}\right)\left(\frac{96500 \text{ C}}{1 \text{ mol e}^-}\right) = 3.26 \text{ X } 10^4 \text{ C}$$

The time that will be required is:

$$\# \text{ hr} = \left(3.26 \text{ X } 10^4 \text{ C}\right)\left(\frac{1 \text{ s}}{1.50 \text{ C}}\right)\left(\frac{1 \text{ hr}}{3600 \text{ s}}\right) = 6.04 \text{ hr}$$

18.29 $Al^{3+}(aq) + 3e^- \rightarrow Al(s)$

The number of Coulombs that are required is:

$$\# C = \left(409 \times 10^3 \text{ g Al}\right)\left(\frac{1 \text{ mol Al}}{26.98 \text{ g Al}}\right)\left(\frac{3 \text{ mol e}^-}{1 \text{ mol Al}}\right)\left(\frac{96500 \text{ C}}{1 \text{ mol e}^-}\right) = 4.39 \times 10^9 \text{ C}$$

The number of amperes is: $4.39 \times 10^9 \text{ C} \div 8.64 \times 10^4 \text{ s} = 5.08 \times 10^4 \text{ A}$

(Note: There are 8.64×10^4 s in 24.0 hr.)

18.31 (a) First, we calculate the number of Coulombs:

$1.50 \text{ A} \times 30.0 \text{ min} \times 60 \text{ s/min} = 2.70 \times 10^3 \text{ A} \cdot \text{s} = 2.70 \times 10^3 \text{ C}$

Then we determine the number of moles of electrons:

$$\# \text{ mol e}^- = \left(2.70 \times 10^3 \text{ C}\right)\left(\frac{1 \text{ mol e}^-}{96500 \text{ C}}\right) = 0.0280 \text{ mol e}^-$$

(b) $0.475 \text{ g} \div 50.9 \text{ g/mol} = 9.33 \times 10^{-3} \text{ mol V}$

(c) $(2.80 \times 10^{-2} \text{ mol e}^-)/(9.33 \times 10^{-3} \text{ mol V}) = 3.00 \text{ mol e}^-/\text{mol V}$

The original oxidation state was V^{3+}.

18.45 Magnesium metal is oxidized at the anode and copper ions are reduced at the cathode to give copper metal: $Mg(s) + Cu^{2+}(aq) \rightarrow Mg^{2+}(aq) + Cu(s)$. The anode half cell is a magnesium wire dipping into a solution of Mg^{2+} ions, and the cathode half cell is a copper wire dipping into a solution of Cu^{2+} ions. Additionally, a salt bridge must connect the two half–cell compartments.

18.48 (a) anode: $Cd(s) \rightarrow Cd^{2+}(aq) + 2e^-$

cathode: $Au^{3+}(aq) + 3e^- \rightarrow Au(s)$

cell: $3Cd(s) + 2Au^{3+}(aq) \rightarrow 3Cd^{2+}(aq) + 2Au(s)$

(b) anode: $Pb(s) + SO_4^{2-}(aq) \rightarrow PbSO_4(s) + 2e^-$

cathode: $PbO_2(s) + SO_4^{2-}(aq) + 4H^+(aq) + 2e^- \rightarrow PbSO_4(s) + 2H_2O(\ell)$

cell: $Pb(s) + PbO_2(s) + 2SO_4^{2-}(aq) + 4H^+(aq) \rightarrow 2PbSO_4(s) + 2H_2O(\ell)$

(c)   anode: $Cr(s) \rightarrow Cr^{3+}(aq) + 3e^-$

        cathode: $Cu^{2+}(aq) + 2e^- \rightarrow Cu(s)$

        cell: $2Cr(s) + 3Cu^{2+}(aq) \rightarrow 2Cr^{3+}(aq) + 3Cu(s)$

18.50 (a) $Fe(s) \,|\, Fe^{2+}(aq) \,||\, Cd^{2+}(aq) \,|\, Cd(s)$

       (b) $Pt(s), Cl^-(aq) \,|\, Cl_2(g) \,||\, Br_2(aq) \,|\, Br^-(aq), Pt(s)$

       (c) $Ag(s) \,|\, Ag^+(aq) \,||\, Au^{3+}(aq) \,|\, Au(s)$

18.60 The difference between the reduction potentials for hydrogen and copper is a constant that is independent of the choice for the reference potential. In other words, the reduction half–cell potential for copper is to be 0.34 units higher for copper than for hydrogen, regardless of the chosen point of reference. If E° for copper is taken to be 0 V, then E° for hydrogen must be –0.34 V.

18.62 (a) $Sn(s)$ (b) $Br^-(aq)$ (c) $Zn(s)$ (d) $I^-(aq)$

18.64 (a) $Fe^{2+}(aq)$ (b) $Ag(s)$ (c) $Ca(s)$ (d) $K(s)$

18.65 (a) $E°_{cell} = -0.40\ V - (-0.44)\ V = 0.04\ V$

       (b) $E°_{cell} = 1.07\ V - (1.36\ V) = -0.29\ V$

       (c) $E°_{cell} = 1.42\ V - (0.80\ V) = 0.62\ V$

18.67 The reactions are spontaneous if the overall cell potential is positive.

      $E°_{cell} = E°_{substance\ reduced} - E°_{substance\ oxidized}$

       (a) $E°_{cell} = 1.42\ V - (0.54\ V) = 0.88\ V$ $\therefore$ spontaneous

       (b) $E°_{cell} = -0.44\ V - (0.96\ V) = -1.40\ V$ $\therefore$ not spontaneous

       (c) $E°_{cell} = -0.74\ V - (-2.76\ V) = 2.02\ V$ $\therefore$ spontaneous

18.70 The given equation is separated into its two half reactions:

$$MnO_4^-(aq) + 8H^+(aq) + 5e^- \rightarrow Mn^{2+}(aq) + 4H_2O(\ell) \qquad \text{reduction}$$

$$5Fe^{2+}(aq) \rightarrow 5Fe^{3+}(aq) + 5e^- \qquad \text{oxidation}$$

      $E°_{cell} = E°_{reduction} - E°_{oxidation} = 1.51\ V - 0.77\ V = 0.74\ V$

18.72 (a) $Zn(s) \,|\, Zn^{2+}(aq) \,||\, Co^{2+}(aq) \,|\, Co(s)$

           $E°_{cell} = -0.28\ V - (-0.76\ V) = 0.48\ V$

(b) $Mg(s) \mid Mg^{2+}(aq) \mid\mid Ni^{2+}(aq) \mid Ni(s)$
$E°_{cell} = -0.25\ V - (-2.37\ V) = 2.12\ V$

(c) $Sn(s) \mid Sn^{2+}(aq) \mid\mid Au^{3+}(aq) \mid Au(s)$
$E°_{cell} = 1.42\ V - (-0.14\ V) = 1.56\ V$

18.74

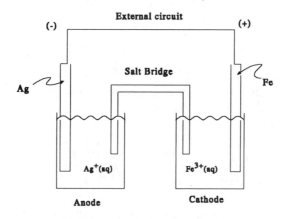

18.76 The half cell with the more positive $E°_{cell}$ will appear as a reduction, and the other half reaction is reversed, to appear as an oxidation:

$BrO_3^-(aq) + 6H^+(aq) + 6e^- \rightarrow Br^-(aq) + 3H_2O$ reduction

$3 \times (2I^-(aq) \rightarrow I_2(s) + 2e^-)$ oxidation

$BrO_3^-(aq) + 6I^-(aq) + 6H^+(aq) \rightarrow 3I_2(s) + Br^-(aq) + 3H_2O$ net reaction

$E°_{cell} = E°_{substance\ reduced} - E°_{substance\ oxidized}$ or
$E°_{cell} = E°_{reduction} - E°_{oxidation} = 1.44\ V - (0.54\ V) = 0.90\ V$

18.78 The half reaction having the more positive standard reduction potential is the one that occurs as a reduction, and the other one is written as an oxidation:

$2 \times (2HOCl(aq) + 2H^+(aq) + 2e^- \rightarrow Cl_2(g) + 2H_2O(\ell))$ reduction

$3H_2O(\ell) + S_2O_3^{2-}(aq) \rightarrow 2H_2SO_3(aq) + 2H^+(aq) + 4e^-$ oxidation

$4HOCl(aq) + 4H^+(aq) + 3H_2O(\ell) + S_2O_3^{2-}(aq) \rightarrow$

$2Cl_2(g) + 4H_2O(\ell) + 2H_2SO_3(aq) + 2H^+(aq)$

which simplifies to give the following net reaction:

$4HOCl(aq) + 2H^+(aq) + S_2O_3^{2-}(aq) \rightarrow 2Cl_2(g) + H_2O(\ell) + 2H_2SO_3(aq)$

Chapter Eighteen

18.80   The two half reactions are:

$$SO_4^{2-}(aq) + 2e^- + 4H^+(aq) \rightarrow H_2SO_3(aq) + H_2O(\ell) \quad \text{reduction}$$

$$2I^-(aq) \rightarrow I_2(s) + 2e^- \qquad\qquad\qquad\qquad \text{oxidation}$$

$$E^\circ{}_{cell} = E^\circ{}_{reduction} - E^\circ{}_{oxidation} = 0.17\ V - (0.54\ V) = -0.37\ V$$

Since the overall cell potential is negative, we conclude that the reaction is not spontaneous in the direction written.

18.84   The maximum amount of work that can be obtained from the cell, per mole of Ag_2O reacting, is given by the absolute value of ΔG° for the reaction, where it must be remembered that the unit volt (V) is equivalent to a joule per Coulomb (J/C):

$$\Delta G^\circ = -n\mathscr{F}E^\circ{}_{cell} = -(2)(96{,}500\ C)(1.50\ J/C) = -2.90 \times 10^5\ J$$

$$\# J = \left(0.500\ g\ Ag_2O\right)\left(\frac{1\ mol\ Ag_2O}{231.7\ g\ Ag_2O}\right)\left(\frac{2.90 \times 10^5\ J}{1\ mol\ Ag_2O}\right) = 626\ J$$

18.86   (a) $E^\circ{}_{cell} = E^\circ{}_{reduction} - E^\circ{}_{oxidation} = 2.01\ V - (1.47\ V) = 0.54\ V$

(b) Since n = 10, $\Delta G^\circ = -n\mathscr{F}E^\circ{}_{cell} = -(10)(96{,}500\ C)(0.54\ J/C) = -5.2 \times 10^5\ J$

$$\Delta G^\circ = -5.2 \times 10^2\ kJ$$

(c) $E^\circ_{cell} = \dfrac{0.0592}{n} \log K_c$

$0.54\ V = (0.0592\ V)/10 \times \log K_c$

$\log K_c = 91.22$ Taking the antilog of both sides of this equation:

$K_c = 1.7 \times 10^{91}$

18.88   First, separate the overall reaction into its two half reactions:

$$2Br^-(aq) \rightarrow Br_2(aq) + 2e^- \quad \text{oxidation}$$

$$I_2(s) + 2e^- \rightarrow 2I^-(aq) \qquad \text{reduction}$$

$$E^\circ{}_{cell} = E^\circ{}_{reduction} - E^\circ{}_{oxidation} = 0.54\ V - (1.07\ V) = -0.53\ V$$

The value of n is 2: $\Delta G^\circ = -n\mathscr{F}E^\circ{}_{cell} = -(2)(96{,}500\ C)(-0.53\ J/C)$

$$= 1.0 \times 10^5\ J = 1.0 \times 10^2\ kJ$$

18.90 Sn is oxidized by two electrons and Ag is reduced by two electrons:

$$E^{\circ}_{cell} = \frac{0.0592}{n} \log K_c$$

$-0.015\ V = 0.0592\ V/2 \times \log K_c$

$\log K_c = -0.51$

$K_c = antilog(-0.51) = 0.31$

18.94 This reaction involves the oxidation of Ag by two electrons and the reduction of Ni by two electrons. The concentration of the hydrogen ion is derived from the pH of the solution: $[H^+] = antilog\ (-pH) = antilog\ (-5) = 1 \times 10^{-5}\ M$

$$E_{cell} = 2.48\ V - \frac{0.0592\ V}{2} \log \frac{\left[Ag^+\right]^2\left[Ni^{2+}\right]}{\left[H^+\right]^4}$$

$$= 2.48\ V - \frac{0.0592\ V}{2} \log \frac{\left[1.0 \times 10^{-2}\right]^2\left[1.0 \times 10^{-2}\right]}{\left[1.0 \times 10^{-5}\right]^4}$$

$E_{cell} = 2.48\ V - 0.41\ V = 2.07\ V$

18.96 The initial numbers of moles of Ag^+ and Zn^{2+} are: $1.00\ mol/L \times 0.100\ L = 0.100\ mol$. The number of Coulombs (A•s) that have been employed is:

$0.10\ C/s \times 15.00\ hr \times 3600\ s/hr = 5.4 \times 10^3\ C$.

The number of moles of electrons is:

$5.4 \times 10^3\ C \div 96,500\ C/mol = 5.6 \times 10^{-2}\ mol\ electrons$.

For Ag^+, there is 1 mol per mole of electrons, and for Zn^{2+}, there are two moles of electrons per mol of Zn. This means that the number of moles of the two ions that have been consumed or formed is given by:

$5.6 \times 10^{-2}\ mol\ e^- \times 1\ mol\ Ag^+/1\ mol\ e^- = 5.6 \times 10^{-2}\ mol\ Ag^+$ reacted.

$5.6 \times 10^{-2}\ mol\ e^- \times 1\ mol\ Zn^{2+}/2\ mol\ e^- = 2.8 \times 10^{-2}\ mol\ Zn^{2+}$ formed

The number of moles of Ag^+ that remain is: $0.100 - 0.056 = 0.044\ mol$ of Ag^+

The final concentration of silver ion is: $[Ag^+] = 0.044\ mol/0.100\ L = 0.44\ M$

The number of moles of Zn^{2+} that are present is: $0.100 + 0.028 = 0.128 \text{ mol } Zn^{2+}$

The final concentration of zinc ion is: $[Zn^{2+}] = 0.128 \text{ mol}/0.100 \text{ L} = 1.28 \text{ M}$

The standard cell potential should be:
$$E^\circ_{cell} = E^\circ_{reduction} - E^\circ_{oxidation} = 0.80 - (-0.76) = 1.56 \text{ V}$$

We now apply the Nernst equation:

$$E_{cell} = E^\circ_{cell} - \frac{0.0592 \text{ V}}{2} \log \frac{1.28}{(0.44)^2}$$

$E_{cell} = 1.56 \text{ V} - 0.024 \text{ V} = 1.54 \text{ V}$

18.98   Since the copper half cell is the cathode, this is the half cell in which reduction takes place. The silver half cell is therefore the anode, where oxidation of silver occurs. The standard cell potential is: $E^\circ_{cell} = E^\circ_{reduction} - E^\circ_{oxidation} = 0.3419 \text{ V} - 0.2223 \text{ V} = 0.1196 \text{ V}$. The overall cell reaction is: $Cu^{2+}(aq) + 2Ag(s) + 2Cl^-(aq) \rightarrow Cu(s) + 2AgCl(s)$, and the Nernst equation becomes:

$$E_{cell} = 0.1196 \text{ V} - \frac{0.0592 \text{ V}}{2} \log \frac{1}{\left[Cu^{2+}\right]\left[Cl^-\right]^2}$$

If we use the various values given in the exercise, we arrive at:
$0.0925 \text{ V} = 0.1196 \text{ V} - 0.0296 \text{ V} \times \log(1/[Cl^-]^2)$, which rearranges to give:
$$\log(1/[Cl^-]^2) = 0.916, \quad [Cl^-] = 0.348 \text{ M}$$

18.100 The half–cell reactions and the overall cell reaction are:
$$Cu^{2+}(aq) + 2e^- \rightarrow Cu(s) \qquad E^\circ_{red} = +0.3419 \text{ V}$$
$$H_2(g) \rightarrow 2H^+(aq) + 2e^- \qquad E^\circ_{ox} = 0.0000 \text{ V}$$
$$Cu^{2+}(aq) + H_2(g) \rightarrow Cu(s) + 2H^+(aq)$$

(a)   The standard cell potential is:
$E°_{cell} = E°_{reduction} - E°_{oxidation} = 0.3419 \text{ V} - 0 \text{ V} = +0.3419 \text{ V}$
The Nernst equation for this system is:

$$E_{cell} = E°_{cell} - \frac{0.0592 \text{ V}}{2} \log \frac{[H^+]^2}{[Cu^{2+}]}$$

which becomes, under the circumstances defined in the problem:

$$E_{cell} = E°_{cell} - \frac{0.0592 \text{ V}}{2} \log [H^+]^2$$

Rearranging the last equation gives:

$$\frac{2 \times (E_{cell} - E°_{cell})}{0.0592 \text{ V}} = -\log [H^+]^2$$

which becomes the desired relationship:

$$\frac{(E_{cell} - E°_{cell})}{0.0592 \text{ V}} = -\log [H^+] = pH$$

(b)   The equation derived in the answer to part (a) of this question is conveniently rearranged to give:

$$E_{cell} = (0.0592 \text{ V})(pH) + E°_{cell} = (0.0592 \text{ V})(5.15) + 0.3419 \text{ V} = 0.647 \text{ V}$$

(c)   The equation that was derived in the answer to part (a) of this question may be used directly:

$$pH = \frac{(E_{cell} - E°_{cell})}{0.0592 \text{ V}} = \frac{(0.645 \text{ V} - 0.3419 \text{ V})}{0.0592 \text{ V}} = 5.12$$

18.110 Reduction of these two metal ions should take place at different applied voltages. First, Ni should "plate out" at –0.25 V, followed by Cd at –0.40 V.

18.112 The cell reaction is:

$$2Cl^-(aq) + 2H_2O(\ell) \rightarrow H_2(g) + Cl_2(g) + 2OH^-(aq).$$

The time is:

20.00 min × 60.00 s/min = 1200 s.

The number of coulombs is:

2.00 A × 1200 s = 2.40×10^3 C.

The number of moles of electrons consumed is:

2.40×10^3 C × 1 mol e^-/96500 C = 0.0249 mol e^-.

Thus the amount of hydroxide produced is:

0.0249 mol e^- × 2 mol OH^-/2 mol e^- = 0.0249 mol OH^-.

Since this will require 0.0249 mol of acid for the titration, the volume required is:

0.0249 mol H^+ ÷ 0.620 M = 0.0402 L = 40.2 mL HCl solution.

18.114 The cathode is positive in a galvanic cell, so we conclude that reduction of platinum ion takes place: $Pt^{2+}(aq) + 2e^- \rightarrow Pt(s)$ $E° = ?$

The anode reaction is: $2\,Ag(s) + 2Cl^-(aq) \rightarrow 2AgCl(s) + 2e^-$ $E° = -0.2223$ V
The overall cell potential is calculated using the Nernst equation:

$$E_{cell} = E°_{cell} - \frac{0.0592 \text{ V}}{2} \log \frac{1}{\left[Pt^{2+}\right]\left[Cl^-\right]^2}$$

$$0.778 \text{ V} = E°_{cell} - \frac{0.0592 \text{ V}}{2} \log \frac{1}{[0.0100][0.100]^2}$$

$E°_{cell}$ = 0.778 V + 0.118 V = 0.896 V

$E°_{Pt^{2+}} - 0.2223$ V = 0.896 V

$E°_{Pt^{2+}}$ = 0.896 V + 0.2223 V = 1.118 V

18.115 $\Delta G° = -nFE° = -\left(2 \text{ mol } e^-\right)\left(\frac{96500 \text{ C}}{1 \text{ mol } e^-}\right)\left(\frac{0.896 \text{ J}}{1 \text{ C}}\right) = -1.73 \times 10^5$ J

19.1 The metal ion is a Lewis acid, accepting a pair of electrons from the ligand, which serves as a Lewis base.

   (a)   The Lewis acid is Cu^{2+}, and the Lewis base is H_2O.

   (b)   The ligand is H_2O.

   (c)   Water provides the donor atom, oxygen.

   (d)   Oxygen is the donor atom, because it is attached to the copper ion.

   (e)   The copper ion is the acceptor.

19.12 The net charge is 1^-, and the formula is $[Co(EDTA)]^-$.

19.14 This is the ion $[Ag(NH_3)_2]^+$, which can exist as the chloride salt: $[Ag(NH_3)_2]Cl$.

19.15 The bonds to NH_3 are known to be stronger, because the test for copper(II) ion requires the ready displacement of water ligands by ammonia ligands to give the ion $[Cu(NH_3)_4]^{2+}$, which has a recognizable deep blue color.

19.28 (a)   The nitrogen atoms are the donor atoms.

   (b)   This is $2 \times 3 = 6$.

   (c)

   (d) $Co(dien)_2^{3+}$, due to the chelate effect

   (e)
   $$H_2-N-CH_2CH_2-NH-CH_2CH_2-NH-CH_2CH_2-NH_2$$

19.32 Since both are the *cis* isomer, they are identical. One can be superimposed on the other after simple rotation.

19.34 No. Neither isomer is chiral since the mirror image may be superimposed on the original.

19.35 (a) The number of moles of chloride that have been precipitated is:

$$\text{\# mol AgCl} = (0.538 \text{ g AgCl})\left(\frac{1 \text{ mol AgCl}}{143.32 \text{ g AgCl}}\right) = 3.75 \times 10^{-3} \text{ mol AgCl}$$

The number of moles of Cr that were originally present is:

$$\text{\# mol Cr} = (0.500 \text{ g CrCl}_3 \cdot 6H_2O)\left(\frac{1 \text{ mol CrCl}_3 \cdot 6H_2O}{266.4 \text{ g CrCl}_3 \cdot 6H_2O}\right)\left(\frac{1 \text{ mol Cr}}{1 \text{ mol CrCl}_3 \cdot 6H_2O}\right)$$

$$= 1.88 \times 10^{-3} \text{ mol Cr}$$

The ratio of moles of Cl^- per mole of Cr is therefore: 3.75/1.88 = 1.99.

This means that there were 2 mol of Cl^- that were free to precipitate. The other mole of chloride ion must have been bound as a ligand to the Cr. In other words, the complex ion was $[Cr(Cl)(H_2O)_5]^{2+}$.

(b) $[Cr(Cl)(H_2O)_5]Cl_2 \cdot H_2O$

(c)

(d) There is only one isomer.

19.37

cis

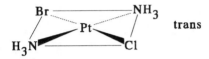

trans

19.39

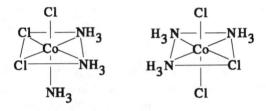

19.42

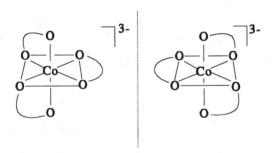

(The curved lines represent C_2O_2 groups.)

19.48

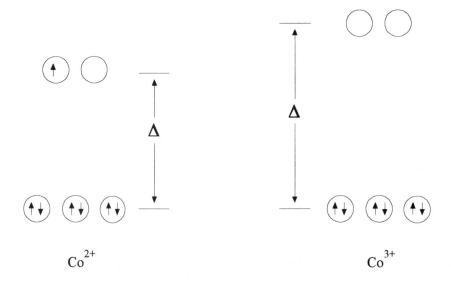

Co^{2+} Co^{3+}

As indicated in the figure above, the oxidation of Co^{2+} to give Co^{3+} removes an electron from a higher energy level. The resulting electron configuration is more stable. Additionally, the energy separation, Δ, increases as the oxidation state increases. This also stabilizes the compound and further explains the ease of oxidation.

19.49 (a) $Cr(H_2O)_6^{3+}$ (b) $Cr(en)_3^{3+}$

19.50 (a) The value of Δ increases down a group. Therefore, we choose: $[Ru(NH_3)_5Cl]^{3+}$
      (b) The value of Δ increases with oxidation state of the metal. Therefore, we choose:
      $[Ru(NH_3)_6]^{3+}$

19.52 $Cr(CN)_6^{3-} < Cr(NO_2)_6^{3-} < Cr(en)_3^{3+} < Cr(NH_3)_6^{3+} < Cr(H_2O)_6^{3+} < CrF_6^{3-} < CrCl_6^{3-}$

19.55 Ligand A produces the larger splitting. The absorbed colors are the complements of the perceived colors:

CoA_6^{3+} – absorbed color is green

CoB_6^{3+} – absorbed color is red

The absorbed color with the highest energy (shortest wavelength) is green. We conclude that ligand A is higher in the spectrochemical series.

19.57 The color of CoA_6^{3+} is red. The complex with the lower oxidation state should have a smaller value for Δ. Therefore, the 2+ ion should absorb in the red, and appear blue.

19.59 $Cr(CN)_6^{3-}$

19.62 This is a weak field complex of Co^{2+}, and it should be a high–spin d^7 case. It cannot be diamagnetic; even if it were low spin, we would still have one unpaired electron.

19.67 (a) $Cu^{2+}(aq) + 4Cl^-(aq) \rightleftharpoons CuCl_4^{2-}(aq)$

$$K_{form} = \frac{\left[CuCl_4^{2-}\right]}{\left[Cu^{2+}\right]\left[Cl^-\right]^4}$$

(b) $Ag^+(aq) + 2I^-(aq) \rightleftharpoons AgI_2^-(aq)$

$$K_{form} = \frac{\left[AgI_2^-\right]}{\left[Ag^+\right]\left[I^-\right]^2}$$

(c) $Cr^{3+}(aq) + 6NH_3(aq) \rightleftharpoons Cr(NH_3)_6^{3+}(aq)$

$$K_{form} = \frac{\left[Cr(NH_3)_6^{3+}\right]}{\left[Cr^{3+}\right]\left[NH_3\right]^6}$$

19.69 $AgCl(s) \rightleftharpoons Ag^+(aq) + Cl^-(aq)$
$Ag^+(aq) + 2NH_3(aq) \rightleftharpoons Ag(NH_3)_2^+(aq)$

Silver chloride is an insoluble solid. However, any Ag^+ ions present react with added NH_3 to form the $Ag(NH_3)_2^+$ complex ion. According to Le Châtelier's Principle, as NH_3 is added to a solution containing Ag^+ ions, the complex ion forms using up the Ag^+ ions. This disrupts the equilibrium and forces AgCl to dissolve.

19.71 (a) $Co(NH_3)_6^{3+}(aq) \rightleftharpoons Co^{3+}(aq) + 6NH_3(aq)$

$$K_{inst} = \frac{\left[Co^{3+}\right]\left[NH_3\right]^6}{\left[Co(NH_3)_6^{3+}\right]}$$

(b) $HgI_4^{2-}(aq) \rightleftharpoons Hg^{2+}(aq) + 4I^-(aq)$

$$K_{inst} = \frac{\left[Hg^{2+}\right]\left[I^-\right]^4}{\left[HgI_4^{2-}\right]}$$

(c) $Fe(CN)_6^{4-}(aq) \rightleftharpoons Fe^{2+}(aq) + 6CN^-(aq)$

$$K_{inst} = \frac{\left[Fe^{2+}\right]\left[CN^-\right]^6}{\left[Fe(CN)_6^{4-}\right]}$$

19.73 (a) $K_{form} = \dfrac{1}{K_{inst}} = \dfrac{1}{5.6 \times 10^{-2}} = 1.8 \times 10^2$

(b) Because this value is small, this ion is much less stable than those listed in the table.

19.75 $AgBr(s) \rightleftharpoons Ag^+ + Br^-$ $K_{sp} = 5.0 \times 10^{-13}$
$Ag^+ + 2S_2O_3^{2-} \rightleftharpoons Ag(S_2O_3)_2^{3-}$ $K_f = 2.0 \times 10^{13}$
$AgBr(s) + 2S_2O_3^{2-} \rightleftharpoons Ag(S_2O_3)_2^{3-} + Br^-$ $K_c = K_{sp}*K_f = 10$

	$[S_2O_3^{2-}]$	$[Ag(S_2O_3)_2^{3-}]$	$[Br^-]$
I	1.20	–	–
C	$-2x$	$+x$	$+x$
E	$1.20-2x$	$+x$	$+x$

Note: Since the AgBr(s) has a constant concentration, it may be neglected.

$$K_c = \frac{[Ag(S_2O_3)_2^{3-}][Br^-]}{[S_2O_3^{2-}]^2} = 10$$

$$K_c = \frac{x^2}{(1.20-2x)^2} = 10$$

To solve this equation, take the square root of both sides and then solve for x.

$x = 0.518 \text{ M} = [\text{Ag}(\text{S}_2\text{O}_3)_2{}^{3-}]$

Since 1 mole of AgBr produces 1 mole of $\text{Ag}(\text{S}_2\text{O}_3)_2{}^{3-}$, we can determine the number of grams of AgBr that will dissolve in 125 mL.

\# g AgBr $= (0.125 \text{ L})(0.518 \text{ moles} / \text{L})(187.77 \text{ g} / \text{mole}) = 12.2 \text{ g AgBr}$

20.4   Equation 20.1 becomes:

$$m = \frac{1.00}{\sqrt{1 - \left(\frac{v}{c}\right)^2}}$$

On substituting the various velocities for the value of v in the above expression, and using the value $c = 3.00 \times 10^8$ m s^{-1}, we get: (a) m = 1.01 kg (b) 3.91 kg (c) 12.3 kg

20.6   Solve the Einstein equation for Δm:

$\Delta m = \Delta E/c^2$

$1 \text{ kJ} = 1.00 \times 10^3 \text{ J} = 1.00 \times 10^3 \text{ kg m}^2 \text{ s}^{-2}$

$\Delta m = 1.00 \times 10^3 \text{ kg m}^2 \text{ s}^{-2} \div (3.00 \times 10^8 \text{ m/s})^2 = 1.11 \times 10^{-14} \text{ kg} = 1.11 \times 10^{-11} \text{ g}$

20.8   The joule is equal to one kg m^2/s^2, and this is employed directly in the Einstein equation: $\Delta m = \Delta E/c^2$, where ΔE is the enthalpy of formation of liquid water, which is available in Table 5.2.

$H_2(g) + 1/2 O_2(g) \rightarrow H_2O(\ell),\quad \Delta H° = -285.9 \text{ kJ/mol}$

$\Delta m = (-285.9 \times 10^3 \text{ kg m}^2/\text{s}^2) \div (3.00 \times 10^8 \text{ m/s}^2) = -3.18 \times 10^{-12} \text{ kg}$

$-3.18 \times 10^{-12} \text{ kg} \times 1000 \text{ g/kg} \times 10^9 \text{ ng/g} = -3.18 \text{ ng}$

The negative value for the mass implies that mass is lost in the reaction.

20.10 The mass of the deuterium nucleus is the mass of the proton (1.00727252 u) plus that of a neutron (1.008665 u), or 2.015938 u. The difference between this calculated value and the observed value is equal to Δm:

$\Delta m = (2.015938 - 2.0135) = 2.4 \times 10^{-3} \text{ u}$

$\Delta E = \Delta mc^2 = (2.4 \times 10^{-3} \text{ u})(1.6606 \times 10^{-27} \text{ kg/u})(3.00 \times 10^8 \text{ m/s})^2$

$\Delta E = 3.6 \times 10^{-13} \text{ kg m}^2/\text{s}^2 = 3.6 \times 10^{-13} \text{ J}$

Since there are two neucleons per deuterium nucleus, we have:

$\Delta E = 3.6 \times 10^{-13} \text{ J/2 nucleons} = 1.8 \times 10^{-13} \text{ J per nucleon}$

20.15 (a) $^{211}_{83}\text{Bi}$ (b) $^{177}_{72}\text{Hf}$ (c) $^{216}_{84}\text{Po}$ (d) $^{19}_{9}\text{F}$

20.17 (a) $^{242}_{94}\text{Pu} \rightarrow ^{4}_{2}\text{He} + ^{238}_{92}\text{U}$

       (b) $^{28}_{12}\text{Mg} \rightarrow ^{0}_{-1}\text{e} + ^{28}_{13}\text{Al}$

       (c) $^{26}_{14}\text{Si} \rightarrow ^{0}_{1}\text{e} + ^{26}_{13}\text{Al}$

       (d) $^{37}_{18}\text{Ar} + ^{0}_{-1}\text{e} \rightarrow ^{37}_{17}\text{Cl}$

20.21 $^{58}_{26}\text{Fe}$

20.25 Barium–140 should have the longer half life because it has an even number of protons and neutrons.

20.30 The neutron-to-proton ratio of lead–164 is too low.

20.34 The more likely process is positron emission, because this produces a product having a higher neutron–to–proton ratio: $^{38}_{19}\text{K} \rightarrow ^{0}_{1}\text{e} + ^{38}_{18}\text{Ar}$.

20.36 Six half life periods correspond to the fraction 1/64 of the initial material. That is, one sixty–fourth of the initial material is left after 6 half lives: 3.00 mg × 1/64 = 0.0469 mg remaining.

20.38 $^{53}_{24}\text{Cr}^{*}$; $^{51}_{23}\text{V} + ^{2}_{1}\text{H} \rightarrow ^{53}_{24}\text{Cr}^{*} \rightarrow ^{1}_{1}\text{p} + ^{52}_{23}\text{V}$.

20.40 $^{80}_{35}\text{Br}$

20.42 $^{55}_{26}\text{Fe}$; $^{55}_{25}\text{Mn} + ^{1}_{1}\text{p} \rightarrow ^{1}_{0}\text{n} + ^{55}_{26}\text{Fe}$.

20.46 This is one curie. It is also: 1.0 Ci × 3.7 X 10^{10} Bq/Ci = 3.7 X 10^{10} Bq

20.53 For a first order process, the rate constant is related to the half life by the equation:
       k = $0.693/t_{1/2}$ Hence, k = 0.693/28.1 yr = 2.47 X 10^{-2} yr^{-1}

       Also, for a first order process, the concentration varies with time according to the equation: $\ln \dfrac{[A]_0}{[A]_t} = kt$

which allows us to solve for the time, t, for the activity to decrease by the specified amount:

$$t = \frac{1}{k} \times \ln\frac{[A]_0}{[A]_t} = \frac{1}{2.47 \times 10^{-2} \text{ yr}^{-1}} \times \ln\frac{\left(0.245 \text{ }^{Ci}/_g\right)}{\left(1.00 \times 10^{-6} \text{ }^{Ci}/_g\right)} = 502 \text{ yr}$$

20.55   The rate constant for the first order process is first determined: $k = 0.693/t_{1/2} = 0.693/6.02$ hr $= 0.115$ hr^{-1}

Also, we know that: $\ln\frac{[A]_0}{[A]_t} = kt$

$$\ln\frac{\left(4.52 \times 10^{-6} \text{ Ci}\right)}{[A]_t} = \left(0.115 \text{ hr}^{-1}\right)\left(8.00 \text{ hr}\right) = 0.920$$

Taking the antiln of both sides of this equation gives:

$$\frac{\left(4.52 \times 10^{-6} \text{ Ci}\right)}{[A]_t} = \text{antiln }(0.920) = 2.51$$

$[A]_t = 1.80 \times 10^{-6}$ Ci, the activity after 8 hours.

20.58   In order to solve this problem, it must be assumed that all of the argon–40 that is found in the rock must have come from the potassium–40, i.e., that the rock contains no other source of argon–40. If the above assumption is valid, then any argon–40 that is found in the rock represents an equivalent amount of potassium–40, since the stoichiometry is 1:1. Since equal amounts of potassium–40 and argon–40 have been found, this indicates that the amount of potassium–40 that remains is exactly half the amount that was present originally. In other words, the potassium–40 has undergone one half life of decay by the time of the analysis. The rock is thus seen to be 1.3×10^9 years old.

20.60 Using equation 20.2 we may determine how long it has been since the tree died.

$$\frac{^{14}C}{^{12}C} = 1.2 \times 10^{-12}\, e^{-t/8270}$$

Taking the natural log we determine:

$$\ln\left(\frac{4.8 \times 10^{-14}}{1.2 \times 10^{-12}}\right) = -t/8270$$

$$t = -8270 \times \ln\left(\frac{4.8 \times 10^{-14}}{1.2 \times 10^{-12}}\right) = 2.7 \times 10^4 \text{ yr}$$

The tree died 2.7×10^4 years ago. This is when the volcanic eruption occurred.

20.63 $^{3}_{1}H + {}^{2}_{1}H \rightarrow {}^{4}_{2}He + {}^{1}_{0}n$.

20.71 $^{235}_{92}U + {}^{1}_{0}n \rightarrow {}^{94}_{38}Sr + {}^{140}_{54}Xe + 2{}^{1}_{0}n$.

20.72 Both will be beta emitters. Both lie above the band of stability, and they can mover closer to it by emitting beta particles.

CHAPTER TWENTY-ONE
Review Exercises

21.1 The inorganic compounds of carbon are generally taken to be the oxides and cyanides of carbon, as well as carbonates and bicarbonates of the metals. Thus, the inorganic compounds in this list are (b), (d), and (e), although (c) CCl_4 might arguably be listed also.

21.4 (a) This is impossible, since there should be three hydrogen atoms attached to the first carbon atom.

(b) This is impossible, since there should be only two hydrogen atoms attached to the first carbon atom.

(c) This is impossible since there should be only one hydrogen atom attached to the third carbon atom.

21.5 (a)
```
        H     H
        |     |
  H —— C —— N —— H
        |
        H
```

(b)
```
        H
        |
  Br —— C —— Br
        |
        H
```

(c)
```
        Cl
        |
  H —— C —— Cl
        |
        Cl
```

(d)
```
        H     H
        |     |
  H —— C —— C —— H
        |     |
        H     H
```

(e)
```
        O
        ‖
  H —— C —— O —— H
```

(f)
```
        O
        ‖
  H —— C —— H
```

(g)
```
        H
        |
  H —— N —— O —— H
```

(h)
```
  H —— C ≡ C —— H
```

(i)
```
        H     H
        |     |
  H —— N —— N —— H
```

(j) $H —— C ≡ N$

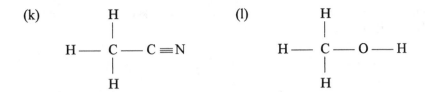

21.9   (a) alkene   (f) alcohol   (k) amine
       (b) alcohol   (g) alkyne   (l) amide
       (c) ester   (h) aldehyde
       (d) carboxylic acid (i) ketone
       (e) amine   (j) ether

21.10   The saturated compounds are b, e, f, j, and k.

21.11   (a) These are identical, being oriented differently only.
       (b) These are identical, being drawn differently only.
       (c) These are unrelated, being alcohols with different numbers of carbon atoms.
       (d) These are isomers, since they have the same empirical formula, but different structures.
       (e) These are identical, being oriented differently only.
       (f) These are identical, being drawn differently only.
       (g) These are isomers, since they have the same empirical formula, but different structures.
       (h) These are identical, being oriented differently only.
       (i) These are isomers, since they have the same empirical formula, but different structures.
       (j) These are isomers, since they have the same empirical formula, but different structures.
       (k) These are identical, being oriented differently only.
       (l) These are identical, being drawn differently only.
       (m) These are isomers, since they have the same empirical formula, but different structures.
       (n) These are unrelated.

21.14   (a)   C
       (b)   A
       (c)   Compound C has two OH groups, and it will have the greatest capacity for hydrogen bonding. Compound A has no polar groups, and has only the very weak London forces.

21.15   Compound B is more soluble in water because of its capacity for hydrogen bonding.

21.17   (a)   pentane
       (b)   2–methylpentane
       (c)   2,4–dimethylhexane
       (d)   2,4–dimethylhexane
       (e)   3–hexene
       (f)   4–methyl–2–pentene

21.20 No, the groups must be different in order to have cis and trans isomers. Isomerism is possible only for (e) and (f).

(e)

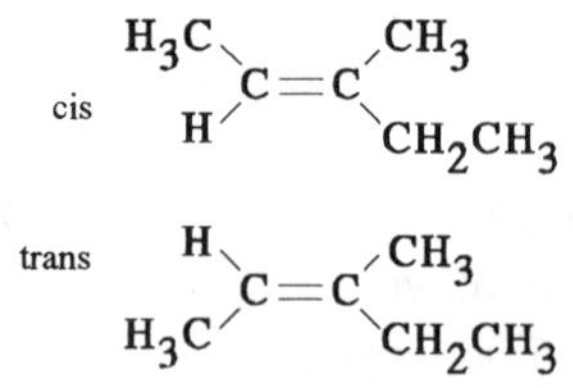

cis

$$H_3C\underset{H}{\diagdown}C=C\overset{CH_3}{\diagup}{CH_2CH_3}$$

trans

$$H\underset{H_3C}{\diagdown}C=C\overset{CH_3}{\diagup}{CH_2CH_3}$$

(f)

cis

$$Cl\underset{CH_3CH_2}{\diagdown}C=C\overset{Cl}{\diagup}{CH_2CH_3}$$

trans

$$Cl\underset{CH_3CH_2}{\diagdown}C=C\overset{CH_2CH_3}{\diagup}{Cl}$$

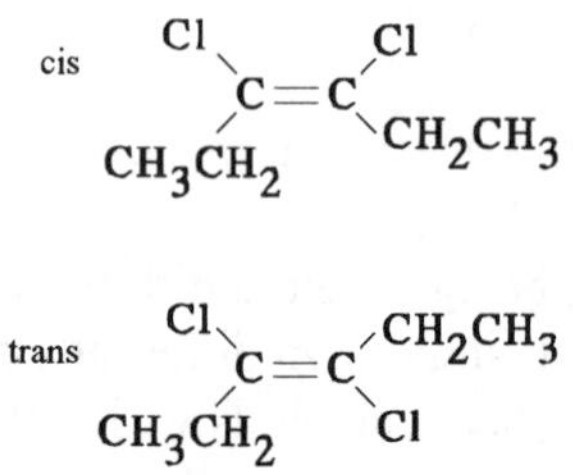

21.23 (a) $CH_3CH_2CH_2CH_3$

(b) $CH_3CH \overset{|}{\underset{|}{—}} CHCH_3$
         Cl   Cl

(c) $CH_3CH \overset{|}{\underset{|}{—}} CHCH_3$
         Br   Br

(d) $CH_3CH \overset{|}{\underset{|}{—}} CHCH_3$
         H   Cl

(e) $CH_3CH \overset{|}{\underset{|}{—}} CHCH_3$
         H   Br

(f) $CH_3CH \overset{|}{\underset{|}{—}} CHCH_3$
         H   OH

21.28
$CH_3CH_2CH_2CH_2OH$	1–butanol
$CH_3CH_2CHOHCH_3$	2–butanol
$(CH_3)_2CHCH_2OH$	2–methyl–1–propanol
$(CH_3)_3COH$	2–methyl–2–propanol

21.31 Those that can be oxidized to aldehydes are:
$CH_3CH_2CH_2CH_2OH$ and $(CH_3)_2CHCH_2OH$

One can be oxidized to a ketone:
$CH_3CH_2CHOHCH_3$

21.32

(a)

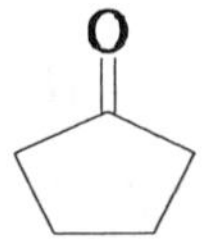

(b)

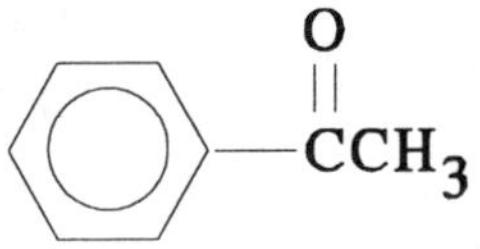

(c)

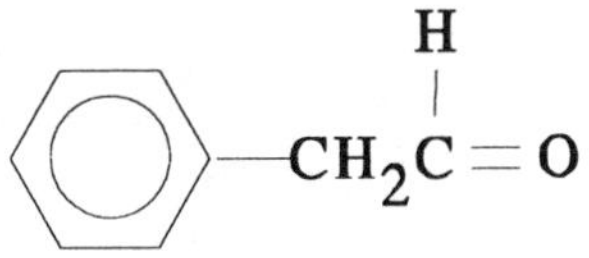

21.36 The elimination of water can result in a C=C double bond in two locations:
$CH_2=CHCH_2CH_3$ $CH_3CH=CHCH_3$

21.39

(a)

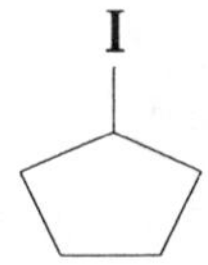

(b)

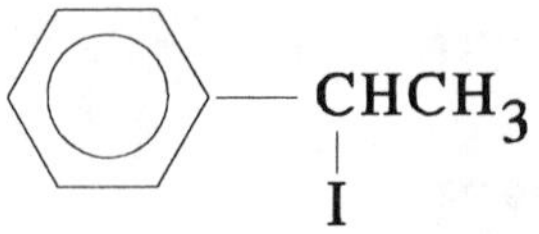

(c)

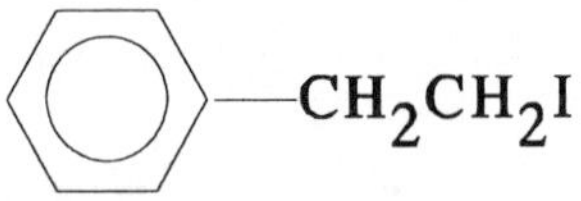

21.42 (a) $CH_3CH_2CH_2NH_3^+$

      (b)   no reaction

      (c)   no reaction

      (d) $CH_3CH_2CH_2NH_2 + H_2O$

21.47 (a)

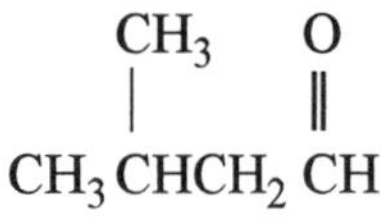

$$\underset{\underset{\displaystyle CH_3}{|}}{CH_3}CHCH_2\,\overset{\overset{\displaystyle O}{\|}}{C}H$$

     (b)

$$CH_3\overset{\overset{\displaystyle O}{\|}}{C}CH_2\,\underset{\underset{\displaystyle CH_3}{|}}{C}HCH_2CH_2CH_2CH_3$$

     (c)

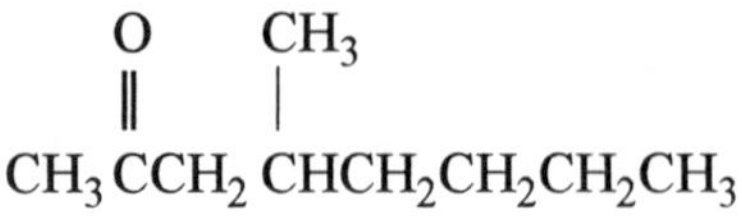

$$CH_3CHClC\overset{\overset{\displaystyle O}{\|}}{}{-}OH$$

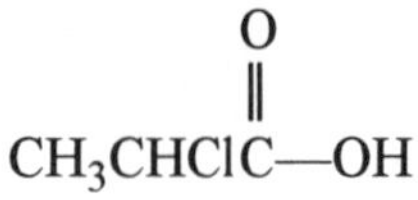

     (d)

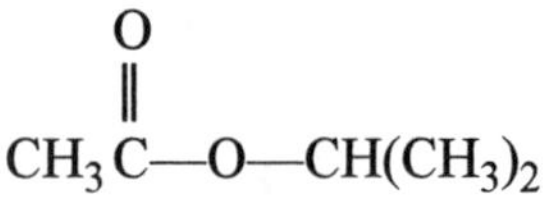

$$CH_3\overset{\overset{\displaystyle O}{\|}}{C}{-}O{-}CH(CH_3)_2$$

     (e)

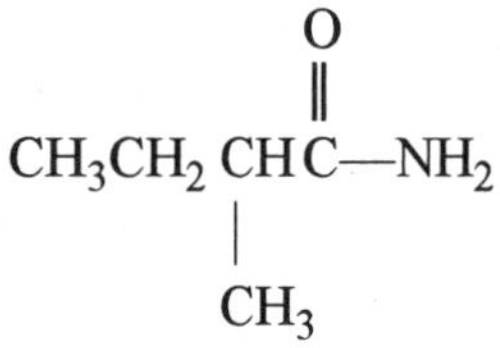

$$CH_3CH_2\,\underset{\underset{\displaystyle CH_3}{|}}{C}H\overset{\overset{\displaystyle O}{\|}}{C}{-}NH_2$$

21.50 (a) This is C, giving:

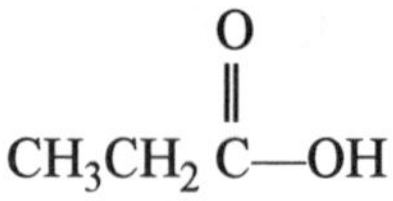

$$CH_3CH_2\,\overset{\overset{\displaystyle O}{\|}}{C}{-}OH$$

     (b) This is B, giving:

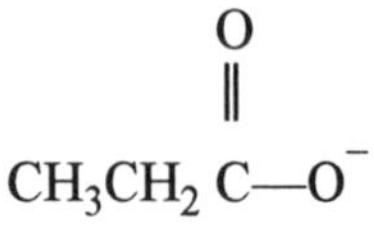

$$CH_3CH_2\,\overset{\overset{\displaystyle O}{\|}}{C}{-}O^-$$

     (c) This is B, giving:

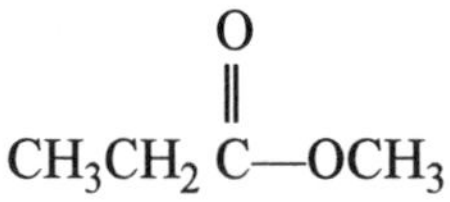

$$CH_3CH_2\,\overset{\overset{\displaystyle O}{\|}}{C}{-}OCH_3$$

     (d) This is A, giving:

$$CH_3CH_2\,\overset{\overset{\displaystyle O}{\|}}{C}{-}H$$

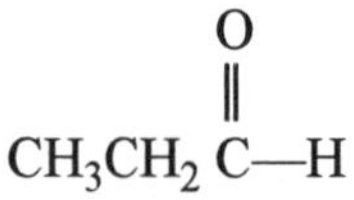

     (e) This is A, giving: $CH_3CH{=}CH_2$

21.51 $CH_3CH_2CO_2H + CH_3OH \rightleftharpoons CH_3CH_2CO_2CH_3 + H_2O$

21.52 (a) $CH_3CH_2CO_2H$

      (b) $CH_3CH_2CO_2H + CH_3OH$

      (c) $Na^+ + CH_3CH_2CH_2CO_2^- + H_2O$

      (d) $CH_3CH_2CONH_2 + H_2O$

      (e) $CH_3CH_2CH_2CO_2H + NH_3$

      (f) $Na^+ + CH_3CH_2CO_2^- + CH_3OH$

21.54 $CH_3CO_2H + CH_3CH_2NHCH_2CH_3$

21.55 (a)   The amine neutralizes HCl:

          $CH_3CH_2CH_2NH_2 + HCl \rightarrow CH_3CH_2CH_2NH_3^+ + Cl^-$

      (b)   The amide is hydrolyzed:

$$CH_3CH_2\overset{\overset{\textstyle O}{\|}}{C}NH_2 + H_2O \rightarrow CH_3CH_2\overset{\overset{\textstyle O}{\|}}{C}OH + NH_3$$

      (c)   The alkylammonium cation neutralizes sodium hydroxide:

          $CH_3CH_2NH_3^+ + OH^- \rightarrow CH_3CH_2NH_2 + H_2O$

21.58

$$\underset{}{-CH_2-\underset{\underset{\textstyle Cl}{|}}{CH}-CH_2-\underset{\underset{\textstyle Cl}{|}}{CH}-CH_2-\underset{\underset{\textstyle Cl}{|}}{CH}-CH_2-\underset{\underset{\textstyle Cl}{|}}{CH}-}$$

$-(-CH_2CHCl-)-$

21.59

$$-O\overset{\overset{\textstyle O}{\|}}{C}CH_2CH_2\overset{\overset{\textstyle O}{\|}}{C}OCH_2CH_2-$$

21.61

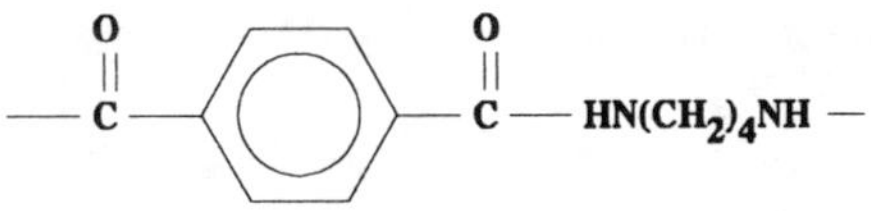

$$-\overset{\overset{\textstyle O}{\|}}{C}-\bigcirc-\overset{\overset{\textstyle O}{\|}}{C}-HN(CH_2)_4NH-$$

21.65   Each yields glucose on complete hydrolysis:

      (a)   glucose   (b)   glucose   (c)   glucose

21.67   The digestion of sucrose yields glucose and fructose.

21.69   They each give glucose:

   (a)   glucose   (b)   glucose

21.75 It is soluble in nonpolar solvents, and it is a substance found in living things.

21.77 No, because it has an odd number of carbon atoms.

21.79 These are, first, the tri-alcohol:

$$\begin{array}{ccc} OH & OH & OH \\ | & | & | \\ CH_2\!\!-\!\!\!\!& CH\!\!-\!\!\!\! & CH_2 \end{array}$$

and the three carboxylic acids:

$$HO\overset{\displaystyle O}{\overset{\|}{C}}(CH_2)_7CH\!\!=\!\!CHCH_2CH\!\!=\!\!CH(CH_2)_4CH_3$$

$$HO\overset{\displaystyle O}{\overset{\|}{C}}(CH_2)_{12}CH_3$$

$$HO\overset{\displaystyle O}{\overset{\|}{C}}(CH_2)_7CH\!\!=\!\!CH(CH_2)_7CH_3$$

21.81 First, we have glycerol:

$$\begin{array}{ccc} OH & OH & OH \\ | & | & | \\ CH_2\!\!-\!\!\!\!& CH\!\!-\!\!\!\! & CH_2 \end{array}$$

and then three anions, as follows.

   (1)   linoleate anion:

$$^-O-\overset{\displaystyle O}{\overset{\|}{C}}(CH_2)_7CH\!\!=\!\!CHCH_2CH\!\!=\!\!CH(CH_2)_4CH_3$$

   (2)   myristate anion:

$$^-O-\overset{\displaystyle O}{\overset{\|}{C}}(CH_2)_{12}CH_3$$

   (3)   oleate anion:

$$^-O-\overset{\displaystyle O}{\overset{\|}{C}}(CH_2)_7CH\!\!=\!\!CH(CH_2)_7CH_3$$

21.86

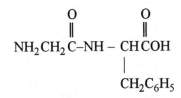

$$\underset{\substack{||\\ O}}{NH_2CH_2\,C}-NHCH_2\,\underset{\substack{||\\ O}}{C}OH$$

21.88

$$\underset{\substack{||\\ O}}{NH_2\,CHC}-NHCH_2\,\underset{\substack{||\\ O}}{C}OH \qquad\qquad \underset{\substack{||\\ O}}{NH_2CH_2\,C}-NH-\underset{\substack{||\\ O}}{CH\,C}OH$$
$$\qquad |\qquad\qquad\qquad\qquad\qquad\qquad\qquad\qquad |$$
$$\quad CH_2C_6H_5\qquad\qquad\qquad\qquad\qquad\qquad CH_2C_6H_5$$

21.100 The genetic messages are carried as a sequence of side chain bases on a DNA segment.

21.102 This is shown in Figures 21.16 and 21.17. Only one particular base can be found opposite another in the double helix. Both structures also have sugar–phosphate backbones.

21.104 (a)   In DNA, A pairs with T.
  (b)   In RNA, A pairs with U.
  (c)   C pairs with G.

21.106 Codons are found on mRNA.

21.107 Anticodons are found on tRNA.

21.109 This is hnRNA (also called ptRNA), which is used in making mRNA.

21.110 Transcription begins with DNA and ends with the synthesis of mRNA.

21.111 Translation begins with mRNA, uses tRNA, and ends with a specific polypeptide.

21.115 (a) $(CH_3)_2CHCH_2OH$ and $(CH_3)_3COH$
  (b)   The one that cannot be oxidized is $(CH_3)_3COH$.
  (c)

$$\underset{+}{CH_3-\underset{\substack{|\\CH_3}}{\overset{\substack{CH_3\\|}}{C}}-CH_3} \qquad\qquad \underset{+}{CH_3-\underset{\substack{|\\H}}{\overset{\substack{CH_3\\|}}{C}}H-CH_2}$$

  (d)   The first one is more stable, since it is the one that leads to the observed product. It is a tertiary carbocation.

21.117 (a) $^-O_2C-C_6H_4-CO_2^-$
  (b) $CH_3CHOHCH_2CH_2CH_3$
  (c) $CH_3NHCH_2CH_2CH_3$
  (d) $CH_3CH_2OCH_2CH_2CO_2H + CH_3OH$

(e) C_5H_8OH

(f) $C_6H_5–CH_2–CHBr–C_6H_5$

21.119 We first find the number of grams of C, H, and O that were present in the original sample:

$$\# \text{ g C } = \left(1.181 \text{ X } 10^{-3} \text{ g CO}_2\right)\left(\frac{12.01 \text{ g C}}{44.01 \text{ g CO}_2}\right) = 3.223 \text{ X } 10^{-4} \text{ g C}$$

$$\# \text{ g H } = \left(0.3653 \text{ X } 10^{-3} \text{ g H}_2\text{O}\right)\left(\frac{2.016 \text{ g H}}{18.02 \text{ g H}_2\text{O}}\right) = 4.087 \text{ X } 10^{-5} \text{ g H}$$

The mass of oxygen is given by:

$$5.574 \text{ X } 10^{-4} \text{ g} - 3.223 \text{ X } 10^{-4} \text{ g} - 4.087 \text{ X } 10^{-5} \text{ g} = 1.942 \text{ X } 10^{-4} \text{ g O}$$

The mole amounts are:

$$3.223 \text{ X } 10^{-4} \text{ g C} \div 12.01 \text{ g/mol} = 2.684 \text{ X } 10^{-5} \text{ mol C}$$

$$4.087 \text{ X } 10^{-5} \text{ g H} \div 1.008 \text{ g/mol} = 4.055 \text{ X } 10^{-5} \text{ mol H}$$

$$1.942 \text{ X } 10^{-4} \text{ g O} \div 16.00 \text{ g/mol} = 1.214 \text{ X } 10^{-5} \text{ mol O}$$

The relative mole amounts are:

for C: $(2.684 \text{ X } 10^{-5} \text{ mol}) \div (1.214 \text{ X } 10^{-5} \text{ mol}) = 2.211$

for H: $(4.055 \text{ X } 10^{-5} \text{ mol}) \div (1.214 \text{ X } 10^{-5} \text{ mol}) = 3.340$

for O: $(1.214 \text{ X } 10^{-5} \text{ mol}) \div (1.214 \text{ X } 10^{-5} \text{ mol}) = 1.000$

These relative mole amounts correspond to the empirical formula:
$C_{20}H_{30}O_9$

CHAPTER ONE
Review Exercises

1.9   (a) $NaH + H_2O \rightarrow NaOH + H_2$

       (b) $CaH_2 + 2H_2O \rightarrow Ca(OH)_2 + 2H_2$

       (c) $HCl(g) + H_2O \rightarrow H_3O^+(aq) + Cl^-(aq)$

       (d) $2Na(s) + H_2(g) \rightarrow 2NaH(s)$

       (e) $Mg(s) + H_2(g) \rightarrow MgH_2(s)$

1.11 MgH_2 – ionic, with considerable covalent character

       H_2Se – molecular

       KH – ionic

       HI – molecular

       PH_3 – molecular

       CaH_2 – ionic

1.20   (a)   The most ionic oxides are those of the Group IA metals.

       (b)   The most basic oxides are those of the Group IA metals.

       (c)   The Group IIA metal oxides (beryllium excepted) are only moderately soluble in water; their basic nature is indicated by the fact that they do dissolve in aqueous acid solutions.

       (d)   The most acidic oxides are the covalent oxides, typically of a nonmetal such as sulfur.

1.23   This is a reaction of a metal oxide with OH^-, and we conclude that the metal oxide acts as an acid. This means that the oxide is an acidic anhydride, or a covalent oxide. The high oxidation state of molybdenum causes the bonds to oxygen to be covalent.

1.26   Peroxides contain an O_2^{2-} group, which may be regarded as derived from the parent peroxide, hydrogen peroxide.

1.28   If each hydrogen in H_2O_2 is assigned the oxidation number +1, then each oxygen is seen to have oxidation state –1.

1.31   (a)   Since this is an acidic solution, we use equation 1.1 for the hydrogen peroxide half reaction:

           reduction $H_2O_2 + 2H^+ + 2e^- \rightarrow 2H_2O$

           We find the half reaction for H_2SO_3 in Table 18.1:

           oxidation $SO_4^{2-} + 4H^+ + 2e^- \rightarrow H_2SO_3 + H_2O$

The net cell reaction is the sum of the above two half reactions:

net cell reaction $\quad H_2O_2 + H_2SO_3 \rightarrow SO_4^{2-} + 2H^+ + H_2O$

(b) $\quad E°_{cell} = E°_{reduction} - E°_{oxidation} = 1.77 - (0.17) = 1.60 \text{ V}$

(c)

$$\# \text{ g } H_2O_2 = (5.25 \text{ g } Na_2SO_3)\left(\frac{1 \text{ mol } Na_2SO_3}{126.1 \text{ g } Na_2SO_3}\right)\left(\frac{1 \text{ mol } SO_3^{2-}}{1 \text{ mol } Na_2SO_3}\right)$$

$$\times \left(\frac{1 \text{ mol } H_2O_2}{1 \text{ mol } SO_3^{2-}}\right)\left(\frac{34.0 \text{ g } H_2O_2}{1 \text{ mol } H_2O_2}\right)$$

$$= 1.41 \text{ g } H_2O_2$$

1.34 (a) $\quad 2NaH(s) + O_2(g) \rightarrow Na_2O(s) + H_2O$

(b) $\quad H^- + H_2O \rightarrow OH^-(aq) + H_2(g)$

(c) $\quad 2HgO(s) \rightarrow 2Hg(\ell) + O_2(g)$

(d) $\quad 2KClO_3(s) \rightarrow 2KCl(s) + 3O_2(g)$

(e) $\quad Na_2O_2(s) + 2H_2O \rightarrow 2NaOH(aq) + H_2O_2(aq)$

(f) $\quad Li(s) + O_2(g) \rightarrow 2Li_2O(s)$

(g) $\quad 2H_2O_2(\ell) \rightarrow O_2(g) + 2H_2O(\ell)$

1.35 (a) $\quad Na_2O(s) + H_2O(\ell) \rightarrow 2NaOH(aq)$

(b) $\quad Na_2O(s) + 2HCl(aq) \rightarrow 2NaCl(aq) + H_2O(\ell)$

(c) $\quad Na_2O(s) + 2HNO_3(aq) \rightarrow 2NaNO_3(aq) + H_2O(\ell)$

(d) $\quad Al_2O_3(s) + 6HCl(aq) \rightarrow 2AlCl_3(aq) + 3H_2O(\ell)$

(e) $\quad Al_2O_3(s) + 2NaOH(aq) + 7H_2O(\ell) \rightarrow 2Na[Al(H_2O)_2(OH)_4](aq)$

(f) $\quad K_2O(s) + 2HCl(aq) \rightarrow 2KCl(aq) + H_2O$

(g) $\quad K_2O(s) + H_2O(\ell) \rightarrow 2KOH(aq)$

(h) $\quad Na_2O_2(s) + 2H_2O \rightarrow 2NaOH(aq) + H_2O_2(aq)$

(i) $\quad Al_2O_3(s) + 6HBr(aq) \rightarrow 2AlBr_3(aq) + 3H_2O(\ell)$

1.40 The smaller value for the heat of vaporization would make ammonia easier to vaporize, and, hence, harder to handle as a liquid.

1.41 (a) $\quad NH_3 + H_2O \rightarrow NH_4^+ + OH^-$

(b) $\quad NH_3(aq) + HCl(aq) \rightarrow NH_4^+(aq) + Cl^-(aq)$

(c) $\quad 4NH_3(g) + 3O_2(g) \rightarrow 2N_2(g) + 6H_2O(g)$

1.42 (a) $NH_3(aq) + HCl(aq) \rightarrow NH_4Cl(aq)$

      (b) $NH_3(aq) + HBr(aq) \rightarrow NH_4Br(aq)$

      (c) $NH_3(aq) + HI(aq) \rightarrow NH_4I(aq)$

      (d) $2NH_3(aq) + H_2SO_4(aq) \rightarrow (NH_4)_2SO_4(aq)$

      (e) $NH_3(aq) + HNO_3(aq) \rightarrow NH_4NO_3(aq)$

1.47 We know that the amide anion is a stronger base than hydroxide, because the amide ion reacts with water to form the hydroxide ion:

$$NH_2^- + H_2O \rightarrow NH_3 + OH^-$$

molecular: $KNH_2(s) + H_2O(\ell) \rightarrow NH_3(aq) + KOH(aq)$

net ionic: $NH_2^- + H_2O \rightarrow NH_3 + OH^-$

1.49 $NH_4^+ + H_2O \rightarrow NH_3 + H_3O^+$

1.52 (a) $Mg_3N_2(s) + 6H_2O(\ell) \rightarrow 3Mg(OH)_2(aq) + 2NH_3(aq)$

      (b) $2Mg(s) + O_2(g) \rightarrow 2MgO(s)$

1.55 These two materials react to give hydrazine, which is toxic.

1.58 These ions have structures based on the addition of one and two protons to the structure of hydrazine itself, as diagramed in the text.

The geometry at each nitrogen atom in $N_2H_6^{2+}$ is likely tetrahedral.

1.60 Ammonia is the stronger base since its value for K_b is larger than that for hydroxylamine.

1.63 From the value of pK_a, we can calculate pK_b for the azide ion: $pK_a + pK_b = 14.00$
pK_b for azide anion is: $14.00 - 4.75 = 9.25$

We conclude that a solution of lithium azide should be basic: $N_3^- + H_2O \rightarrow HN_3 + OH^-$

1.73   The decomposition of N_2O produces more moles of gaseous products than there are moles of gas from the reactant: $2N_2O \rightarrow 2N_2 + O_2$

1.74   The dimerization of NO_2 to give N_2O_4 causes the pairing of two odd electrons.

1.76   (a) $k_a = \text{antilog}(-pK_a) = \text{antilog}(-3.35) = 4.5 \times 10^{-4}$

         (b) $HNO_2(aq) = H^+(aq) + NO_2^-(aq)$

1.77 $NaNO_2$ is a weak base, which causes its aqueous solutions to have a pH greater than 7:

$$NO_2^-(aq) + H_2O(\ell) = HNO_2(aq) + OH^-(aq)$$

1.93 $2H^+(aq) + CO_3^{2-}(aq) \rightarrow H_2O(\ell) + CO_2(g)$

       $H^+(aq) + HCO_3^-(aq) \rightarrow H_2O(\ell) + CO_2(g)$

1.95 $CN^-(aq) + H_2O(\ell) \rightarrow HCN(aq) + OH^-(aq)$

1.101   (a) $Zn(s) + 2HCl(aq) \rightarrow ZnCl_2(aq) + H_2(g)$

         (b) $Ca(s) + H_2(g) \rightarrow CaH_2(s)$

         (c) $NaH(s) + H_2O(\ell) \rightarrow NaOH(aq) + H_2(g)$

         (d) $Mg(s) + 2HCl(aq) \rightarrow MgCl_2(aq) + H_2(g)$

         (e) $KH(s) + H_2O(\ell) \rightarrow KOH(aq) + H_2(g)$

         (f) $CH_3-CH=CH_2(g) + H_2(g) \rightarrow CH_3-CH_2-CH_3(g)$

         (g) $CH_2=CH_2(g) + CO(g) + H_2(g) \rightarrow CH_3-CH_2-CH=O(\ell)$

         (h) $2HgO(s) \rightarrow 2Hg(\ell) + O_2(g)$

         (i) $Na_2O(s) + H_2O(\ell) \rightarrow 2NaOH(aq)$

         (j) $MgO(s) + 2HBr(aq) \rightarrow MgBr_2(aq) + H_2O(\ell)$

         (k) $2KClO_3(s) \rightarrow 2KCl(s) + 3O_2(g)$

         (l) $Al_2O_3(s) + 6HBr(aq) \rightarrow 2AlBr_3(aq) + 3H_2O(\ell)$

         (m) $Al_2O_3(s) + 2KOH(aq) + 7H_2O(\ell) \rightarrow 2K[Al(H_2O)_2(OH)_4](aq)$

         (n) $2CrO_3(s) + H_2O(\ell) \rightarrow H_2CrO_4(aq)$

         (o) $2H_2O_2(aq) \rightarrow 2H_2O(\ell) + O_2(g)$

         (p) $K_2O_2(s) + 2H_2O(\ell) \rightarrow 2KOH(aq) + H_2O_2(aq)$

         (q) $CO_2(aq) + H_2O \rightarrow H_2CO_3(aq)$

         (r) $CO_2(aq) + OH^- \rightarrow HCO_3^-$

         (s) $CaC_2 + 2H_2O \rightarrow C_2H_2(g) + Ca(OH)_2$

(t) $Mg_2C(s) + 2H_2O \rightarrow C_2H_2(g) + 2Mg(OH)_2(s)$

(u) $CaCO_3(s) \rightarrow CaO(s) + CO_2(g)$

(v) $NaCN(aq) + HCl(aq) \rightarrow NaCl(aq) + HCN(g)$

1.104 (a) $N_2(g) + 2H_2(g) \rightarrow N_2H_4(\ell)$, $\Delta H° = +50.6$ kJ

(b)   It is unstable, since it is formed endothermically.

(c)   The activation energy for the reaction is so high that the rate of reaction is low.

(d) $\Delta G° = \Delta H° - T\Delta S°$

149.2 kJ = 50.6 kJ − (298 K)($\Delta S°$)

$\Delta S° = -0.331$ kJ K^{-1} mol^{-1} = −331 J K^{-1} mol^{-1}

Since the value of $\Delta S°$ is negative, the reaction is not spontaneous from the standpoint of entropy. This is because, on going from reactants to products, there is a decrease in the number of molecules of gas, and, hence, a decrease in disorder.

1.106 (a)   oxidation: $NO + 2H_2O \rightarrow NO_3^- + 4H^+ + 3e^-$, $E° = 0.96$ V

reduction: $NO_3^- + 2H^+ + e^- \rightarrow NO_2 + H_2O$, $E° = +0.80$ V

(b)   Add the two equations given in the answer to part (a):

$2NO_3^- + NO + 2H^+ \rightarrow 3NO_2 + H_2O$

(c) $E°_{cell} = E°_{reduction} - E°_{oxidation} = 0.80 - (0.96) = -0.04$ V

The negative value of E° indicates that the reaction is not spontaneous under standard conditions.

(d)   The value of $E°_{cell}$ is only slightly negative, and the second term in the Nernst equation can be made to be sufficiently positive so as to offset the negative value of $E°_{cell}$.

1.108 (a) $HCl(aq) + NaOH(aq) \rightarrow H_2O(\ell) + NaCl(aq)$

$H^+(aq) + OH^-(aq) \rightarrow H_2O(\ell)$

(b) $HCl(aq) + NaHCO_3(aq) \rightarrow H_2O(\ell) + CO_2(g) + NaCl(aq)$

$H^+(aq) + HCO_3^-(aq) \rightarrow H_2O(\ell) + CO_2(g)$

(c) $2HCl(aq) + Na_2CO_3(aq) \rightarrow H_2O(\ell) + CO_2(g) + 2NaCl(aq)$

$2H^+(aq) + CO_3^{2-}(aq) \rightarrow H_2O(\ell) + CO_2(g)$

(d) $HCl(aq) + KOH(aq) \rightarrow H_2O(\ell) + KCl(aq)$

$H^+(aq) + OH^-(aq) \rightarrow H_2O(\ell)$

(e) $2HCl(aq) + K_2CO_3(aq) \rightarrow H_2O(\ell) + CO_2(g) + 2KCl(aq)$

$2H^+(aq) + CO_3^{2-}(aq) \rightarrow H_2O(\ell) + CO_2(g)$

(f) $HCl(aq) + KHCO_3(aq) \rightarrow H_2O(\ell) + CO_2(g) + KCl(aq)$

$H^+(aq) + HCO_3^-(aq) \rightarrow H_2O(\ell) + CO_2(g)$

(g) $2HCl(aq) + CaCO_3(s) \rightarrow H_2O(\ell) + CO_2(g) + CaCl_2(aq)$

$2H^+(aq) + CaCO_3(s) \rightarrow H_2O(\ell) + CO_2(g) + Ca^{2+}(aq)$

(h) $2HCl(aq) + Ca(OH)_2(s) \rightarrow 2H_2O(\ell) + CaCl_2(aq)$

$2H^+(aq) + Ca(OH)_2(s) \rightarrow 2H_2O(\ell) + Ca^{2+}(aq)$

(i) $2HCl(aq) + Mg(OH)_2(s) \rightarrow 2H_2O(\ell) + MgCl_2(aq)$

$2H^+(aq) + Mg(OH)_2(s) \rightarrow 2H_2O(\ell) + Mg^{2+}(aq)$

(j) $2HCl(aq) + MgCO_3(s) \rightarrow H_2O(\ell) + CO_2(g) + MgCl_2(aq)$

$2H^+(aq) + MgCO_3(s) \rightarrow H_2O(\ell) + CO_2(g) + Mg^{2+}(aq)$

(k) $2HCl(aq) + Na_2S(aq) \rightarrow 2NaCl(aq) + H_2S(g)$

$2H^+(aq) + S^{2-}(aq) \rightarrow H_2S(g)$

(l) $2HCl(aq) + K_2SO_3(aq) \rightarrow 2KCl(aq) + SO_2(g) + H_2O$

$2H^+(aq) + SO_3^{2-}(aq) \rightarrow SO_2(g) + H_2O$

(m)   no reaction

(n)   no reaction

(o) $HCl(aq) + LiCN(aq) \rightarrow HCN(g) + LiCl(aq)$

$H^+(aq) + CN^-(aq) \rightarrow HCN(g)$

(p) $2HCl(aq) + Pb(NO_3)_2(aq) \rightarrow PbCl_2(s) + 2HNO_3$

$2Cl^-(aq) + Pb^{2+}(aq) \rightarrow PbCl_2(s)$

(q) $HCl(aq) + AgNO_3(aq) \rightarrow AgCl(s) + HNO_3(aq)$

$Cl^-(aq) + Ag^+(aq) \rightarrow AgCl(s)$

(r) $2HCl(aq) + Ca(s) \rightarrow H_2(g) + CaCl_2(aq)$

$2H^+(aq) + Ca(s) \rightarrow Ca^{2+}(aq) + H_2(g)$

(s) $HCl(aq) + NaNO_2(aq) \rightarrow HNO_2(aq) + NaCl(aq)$

$H^+(aq) + NO_2^-(aq) \rightarrow HNO_2(aq)$

(t)   no reaction

(u) $2HCl(aq) + Mg(s) \rightarrow MgCl_2(aq) + H_2(g)$

$2H^+(aq) + Mg(s) \rightarrow Mg^{2+}(aq) + H_2(g)$

(v) $HCl(aq) + NaC_2H_3O_2(aq) \rightarrow NaCl(aq) + HC_2H_3O_2(aq)$

$H^+(aq) + C_2H_3O_2^-(aq) \rightarrow HC_2H_3O_2(aq)$

(w) $HCl(aq) + NaNH_2(s) \rightarrow NaCl(aq) + NH_3(aq)$

$H^+(aq) + NaNH_2(s) \rightarrow Na^+(aq) + NH_3(aq)$

(x) $HCl(aq) + KN_3(s) \rightarrow KCl(aq) + HN_3(aq)$

$H^+(aq) + KN_3(s) \rightarrow HN_3(aq) + K^+(aq)$

(y) $HCl(aq) + NH_3(aq) \rightarrow NH_4Cl(aq)$

$H^+(aq) + NH_3(aq) \rightarrow NH_4^+(aq)$

(z)   no reaction

CHAPTER TWO
Review Exercises

2.10   This is termed sulfurous acid, $H_2SO_3(aq)$, although H_2SO_3 molecules have never been found. Such solutions contain hydrated sulfur dioxide, $SO_2 \cdot nH_2O$, $H^+(aq)$, and HSO_3^- (aq).

2.11 (a) $NaOH(aq) + SO_2(g) \rightarrow NaHSO_3(aq)$

      (b) $NaHCO_3(aq) + SO_2(g) \rightarrow NaHSO_3(aq) + CO_2(g)$

2.14 $I_2(s) + H_2SO_3(aq) + H_2O(\ell) \rightarrow 2I^-(aq) + SO_4^{2-}(aq) + 4H^+(aq)$

$E°_{cell} = E°_{red} - E°_{ox}$, and the standard half–cell potentials are obtained from Table 18.1.

$E°_{cell} = 0.54 - (0.17) = 0.37$ V

The reaction is spontaneous in the direction written, because the overall cell potential is greater than zero.

2.16 (a) $NaHSO_3(aq) + HCl(aq) \rightarrow H_2O(\ell) + SO_2(g) + NaCl(aq)$

$HSO_3^-(aq) + H^+(aq) \rightarrow H_2O(\ell) + SO_2(g)$

      (b) $Na_2SO_3(aq) + 2HCl(aq) \rightarrow H_2O(\ell) + SO_2(g) + 2NaCl(aq)$

$SO_3^{2-}(aq) + 2H^+(aq) \rightarrow SO_2(g) + H_2O(\ell)$

2.18 (a) $SO_3(g) + H_2O(\ell) \rightarrow H_2SO_4(aq)$

      (b) $SO_3(g) + 2NaHCO_3(aq) \rightarrow Na_2SO_4(aq) + H_2O(\ell) + 2CO_2(g)$

      (c) $SO_3(g) + Na_2CO_3(aq) \rightarrow Na_2SO_4(aq) + CO_2(g)$

      (d) $SO_3(g) + 2NaOH(aq) \rightarrow Na_2SO_4(aq) + H_2O(\ell)$

2.23   The solid was the acid $NaHSO_4$, reacting with sodium carbonate to release $CO_2(g)$.
$$2NaHSO_4(aq) + Na_2CO_3(s) \rightarrow 2Na_2SO_4(aq) + H_2O(\ell) + CO_2(g)$$

2.30   (a) $S_2O_3^{2-}(aq) + 4Cl_2(g) + 5H_2O(\ell) \rightarrow 2HSO_4^-(aq) + 8H^+(aq) + 8Cl^-(aq)$

        (b) $2S_2O_3^{2-}(aq) + I_2(aq) \rightarrow S_4O_6^{2-}(aq) + 2I^-(aq)$

2.34 $HS^- \rightleftharpoons H^+ + S^{2-}$
$$1.3 \times 10^{-13} = x^2/(0.100 - x)$$
$$x = 1.1 \times 10^{-7} \text{ M}$$
$$pH = -\log(1.1 \times 10^{-7}) = 6.96$$

2.40 $H_2SeO_4(aq) + 2NaOH(aq) \rightarrow Na_2SeO_4(aq) + 2H_2O(\ell)$

2.47   (a)   In the first case, we have the formation of phosphorous acid:
$$PCl_3(\ell) + 3H_2O(\ell) \rightarrow H_3PO_3(aq) + 3HCl(aq)$$
        (b)   In the second case, we have the formation of phosphoric acid:
$$PCl_5(\ell) + 4H_2O(\ell) \rightarrow H_3PO_4(aq) + 5HCl(aq)$$

2.48 $PBr_3(\ell) + 3H_2O(\ell) \rightarrow H_3PO_3(aq) + 3HBr(aq)$
       $PBr_5(s) + 4H_2O(\ell) \rightarrow H_3PO_4(aq) + 5HBr(aq)$

2.57   (a)   arsenic acid
        (b)   sodium arsenate
        (c)   arsenous acid
        (d)   sodium dihydrogen arsenate
        (e)   arsenic trichloride
        (f)   antimony pentachloride

2.60 $2F_2(g) + 2H_2O(\ell) \rightarrow 4HF(g) + O_2(g)$
       $Cl_2(aq) + H_2O(\ell) \rightarrow HCl(aq) + HOCl(aq)$
       $Br_2(aq) + H_2O(\ell) \rightarrow HBr(aq) + HOBr(aq)$
       $I_2(aq) + H_2O(\ell) \rightarrow HI(aq) + HOI(aq)$

2.66   This is the oxidation of iodide ion by oxygen:
$$O_2(g) + 4HI(aq) \rightarrow 2I_2(s) + 2H_2O(\ell)$$

2.68   (a) Y^- is oxidized.

        (b) X_2 is reduced.

        (c)   No. Cl_2 is known to oxidize Br^- instead.

        (d)   Yes. Cl_2 is a stronger oxidizing agent than Br_2.

        (e)   Yes. Br_2 is a stronger oxidizing agent than I_2.

2.79   (a)   First, calculate the amount of phosphorus that was present in the original compound:

$$\# \, g\,P = \left(0.5798 \text{ g Ba}_3(PO_4)_2\right)\left(\frac{1 \text{ mol Ba}_3(PO_4)_2}{601.92 \text{ g Ba}_3(PO_4)_2}\right)$$

$$\times \left(\frac{2 \text{ mol P}}{1 \text{ mol Ba}_3(PO_4)_2}\right)\left(\frac{30.97 \text{ g P}}{1 \text{ mol P}}\right)$$

$$= 0.05966 \text{ g P}$$

The mass of sulfur that was present in the original compound is therefore:
$$0.2141 \text{ g} - 0.05966 \text{ g} = 0.1544 \text{ g S}$$
The mass percentages are:
$$\%P = (0.05966 \text{ g}/0.2141 \text{ g}) \times 100 = 27.87 \, \% \text{ P}$$
$$\%S = (0.1544 \text{ g}/0.2141 \text{ g}) \times 100 = 72.12 \, \% \text{ S}$$

In 100 g of the original compound, we therefore have the following number of moles of each:

$$\text{mol P} = 27.87 \text{ g}/30.97 \text{ g/mol} = 0.8999 \text{ mol P}$$
$$\text{mol S} = 72.12 \text{ g}/32.06 \text{ g/mol} = 2.250 \text{ mol S}$$

The relative mole amount of these two elements is:
     for P, 0.8999 mol/0.8999 mol = 1.000 mol P
     for S, 2.250 mol/0.8999 mol = 2.500 mol S

The empirical formula is P_2S_5.

        (b)   The mass of the empirical unit is:
$$2 \times (30.97) + 5 \times (32.06) = 222.2 \text{ g}$$
Since this is half the molecular weight, the molecular formula is P_4S_{10}.

        (c)   The gas is H_2S.

2.80 (a) $IO_3^- + 3HSO_3^- \rightarrow I^- + 3SO_4^{2-} + 3H^+$

(b) $5I^- + IO_3^- + 6H^+ \rightarrow 3I_2 + 3H_2O$

(c) A 6.00 % composition indicates that, in 100.0 g of solution, we have 6.00 g of iodate, which is present as the sodium salt.

$$\# \text{ g NaHSO}_3 = (6.00 \text{ g NaIO}_3)\left(\frac{1 \text{ mol NaIO}_3}{197.9 \text{ NaIO}_3}\right)$$

$$\times \left(\frac{3 \text{ mol NaHSO}_3}{1 \text{ mol NaIO}_3}\right)\left(\frac{104.1 \text{ g NaHSO}_3}{1 \text{ mol NaHSO}_3}\right)$$

$$= 9.47 \text{ g NaHSO}_3$$

(d) The answer to this question may be broken into three steps:

1) Determine the amount of I^- formed in reaction (a)

2) Determine the amount of $NaIO_3$ that will react with the I^-.

3) Determine the amount of solution needed to obtain this amount of $NaIO_3$.

$$\# \text{ mol } I^- = (6.00 \text{ g NaIO}_3)\left(\frac{1 \text{ mol NaIO}_3}{197.9 \text{ NaIO}_3}\right)$$

$$\times \left(\frac{1 \text{ mol IO}_3^-}{1 \text{ mol NaIO}_3}\right)\left(\frac{1 \text{ mol } I^-}{1 \text{ mol IO}_3^-}\right)$$

$$= 0.0304 \text{ mol } I^-$$

$$\# \text{ g NaIO}_3 = (0.0304 \text{ mol } I^-)\left(\frac{1 \text{ mol IO}_3^-}{5 \text{ mol } I^-}\right)\left(\frac{1 \text{ mol NaIO}_3}{1 \text{ mol IO}_3^-}\right)\left(\frac{197.9 \text{ g NaIO}_3}{1 \text{ mol NaIO}_3}\right)$$

$$= 1.20 \text{ g NaIO}_3$$

This must be obtained from a solution that is 6.00 % by mass. The mass of solution that is needed is therefore: 6.00 % of X = 1.20 g
X = 20.0 g of brine solution

2.81 The data to solve this problem is found in Section 2.3. Specifically, the subsection entitled "Oxides and Oxoacids of Chlorine."

$$K = \frac{[HCl][HOCl]}{[Cl_2]} = \frac{(0.030)(0.030)}{(0.061)} = 0.015$$

CHAPTER THREE
Review Exercises

3.3   (a)   Tl   (b)   Ba   (c)   K

3.5   MgO

3.9   There is no economical or practical way of recovering aluminum metal from this material.

3.43   (a) $2Li + 2H_2O \rightarrow 2LiOH + H_2$

      (b) $2K + 2H_2O \rightarrow 2KOH + H_2$

      (c) $2Rb + 2H_2O \rightarrow 2RbOH + H_2$

3.45   (a) $Na_2O_2 + 2H_2O \rightarrow 2NaOH + H_2O_2$

      (b) $Na_2O + H_2O \rightarrow 2Na^+ + 2OH^-$

      (c) $2KO_2 + H_2O \rightarrow 2K^+ + OH^- + HO_2^- + O_2$

      (d) $NaOH + CO_2 \rightarrow NaHCO_3$

3.53 $CO_3^{2-} + H_2O \rightleftharpoons HCO_3^- + OH^-$, $K_b = 2.1 \times 10^{-4}$

      $K_b = 2.1 \times 10^{-4} = x^2/1.0$ $\therefore$ $x = 1.5 \times 10^{-2}$ M

      $pOH = -\log [OH^-] = -\log (1.5 \times 10^{-2}) = 1.84$
      $pH = 14.00 - pOH = 14.00 - 1.84 = 12.16$

3.54   (a) $NaClO_4$ (b) $KC_2H_3O_2$ (c) $KC_2H_3O_2$

3.57

$$\text{\# lbs Na}_2\text{CO}_3 = \left(20000 \text{ gal HNO}_3\right)\left(\frac{3.79 \text{ L}}{1 \text{ gal}}\right)\left(\frac{16 \text{ mol HNO}_3}{1 \text{ L HNO}_3}\right)$$

$$\times \left(\frac{1 \text{ mol Na}_2\text{CO}_3}{2 \text{ mol HNO3}}\right)\left(\frac{106 \text{ g Na}_2\text{CO}_3}{1 \text{ mol Na}_2\text{CO}_3}\right)\left(\frac{1 \text{ lb}}{453.6 \text{ g}}\right)$$

$$= 1.42 \times 10^5 \text{ lbs Na}_2\text{CO}_3$$

3.77 $Mg(OH)_2$
       $Mg(OH)_2 + 2HCl \rightarrow MgCl_2 + 2H_2O$

3.80 $2Mg + O_2 \rightarrow 2MgO$
$MgCO_3 \rightarrow MgO + CO_2$
$Mg(OH)_2 \rightarrow MgO + H_2O$

3.116 PbO_2 is a better oxidizing agent than SnO_2.

3.121 $Pb(OH)_4^{2-} + ClO^- \rightarrow PbO_2 + Cl^- + 2OH^- + H_2O$

3.122 $4H^+ + PbO_2 + 4Cl^- \rightarrow PbCl_2 + 2H_2O + Cl_2$

3.126 In each case, $K_{sp} = 4x^3$, where x is the molar solubility.
   (a) $1.7 \times 10^{-5} = 4x^3$, $\therefore x = 1.6 \times 10^{-2}$ M for $PbCl_2$
   (b) $2.1 \times 10^{-6} = 4x^3$, $\therefore x = 8.1 \times 10^{-3}$ M for $PbBr_2$
   (c) $7.9 \times 10^{-9} = 4x^3$, $\therefore x = 1.3 \times 10^{-3}$ M for PbI_2

3.130 This is lead chromate, $PbCrO_4$.

3.142 If the atoms have roughly the same size, then the ratio of their atomic masses should be equal to the ratio of their densities. The mass ratio is $180.95 \div 92.91 = 1.948$, so we estimate that the density of Ta should be nearly twice that of Nb, or 16.7 g/cm^3.

3.147 The magnetic domains that are responsible for the ferromagnetic effect are disrupted when the atoms are made to adopt more random arrangements. This allows the domains to become aligned with the earth's magnetic field. See Figure 3.29.

3.154 A water molecule that is attached to the Bi^{3+} center finds its O–H bonds polarized by the bismuth ion, and loss of two protons from this water ligand gives the O atom of BiO^+:
$$Bi–OH_2^{3+} \rightarrow Bi–O^+ + 2H^+$$

3.162 $Cr(H_2O)_6^{3+}(aq) + H_2O(\ell) \rightleftharpoons Cr(H_2O)_5(OH)^{2+}(aq) + H_3O^+(aq)$

3.166 $H_2CrO_4 \rightleftharpoons H^+ + HCrO_4^-$ and $HCrO_4^- \rightleftharpoons H^+ + CrO_4^{2-}$

3.173 (a) $Mn(s) + 2HCl(aq) \rightarrow MnCl_2(aq) + H_2(g)$
   (b) $MnCl_2(aq) + 2NaOH(aq) \rightarrow Mn(OH)_2(s) + 2NaCl(aq)$

3.187 $Fe(s) + 2H^+(aq) \rightarrow Fe^{2+}(aq) + H_2(g)$

3.193 The iron derivative does not dissolve in excess hydroxide solutions, but the chromium derivative does. This is because the iron compound is not acidic; it is more properly written $Fe_2O_3 \cdot xH_2O$, i.e. as the hydrous oxide. As a metal oxide, it is basic.

3.194 Both ions cause hydrolysis of one or more attached water groups, but the Fe^{3+} ion does this more strongly. This is because the Fe^{3+} ion has a higher charge and a smaller size, i.e. a greater charge density. It therefore causes a greater polarization of the O–H bond in the attached water ligands.

3.198 The formation of Prussian Blue from aqueous cyanide solutions can be used for this test. Both Fe^{2+} and Fe^{3+} should be added.

3.203 $4Co^{3+}(aq) + 2H_2O(\ell) \rightarrow 4Co^{2+}(aq) + O_2(g) + 4H^+(aq)$

3.216 Gold leaf is composed of very thin sheets of metallic gold. From the text, we know that 11,000 leaflets equals 1 mm. Therefore, the thickness of one leaf is:

$$\# \text{ mm} = \frac{1 \text{ mm}}{11{,}000 \text{ leafs}} = 9.09 \times 10^{-5} \text{ mm / leaf}$$

$$\# \text{ in.} = \left(9.09 \times 10^{-5} \text{ mm / leaf}\right)\left(\frac{1 \text{ in.}}{25.4 \text{ mm}}\right) = 3.58 \times 10^{-6} \text{ in./leaf}$$

3.220 (a) no reaction
       (b) $Cu + 2H_2SO_4 \rightarrow CuSO_4 + SO_2 + 2H_2O$
       (c) $3Cu + 8HNO_3 \rightarrow 3Cu(NO_3)_2 + 2NO + 4H_2O$
       (d) $Cu + 4HNO_3 \rightarrow Cu(NO_3)_2 + 2NO_2 + 2H_2O$

3.225 $Cu^{2+}(aq) + 2OH^-(aq) \rightarrow Cu(OH)_2(s)$

$Cu(OH)_2(s) + 2OH^-(aq) \rightarrow Cu(OH)_4^{2-}(aq)$

3.226 $Cu(CN)_4^{2-}$

3.228 $Ag + 2H^+ + NO_3^- \rightarrow Ag^+ + NO_2 + H_2O$

$3Ag + 4H^+ + NO_3^- \rightarrow 3Ag^+ + NO + 2H_2O$

3.233 Gold(III) is readily reduced.

3.238 $Zn^{2+}(aq) + 2OH^-(aq) \rightarrow Zn(OH)_2(s)$

$Zn(OH)_2(s) + 2OH^-(aq) \rightarrow Zn(OH)_4^{2-}(aq)$

3.241 $4Zn(s) + NO_3^-(aq) + 10H^+(aq) \rightarrow 4Zn^{2+}(aq) + NH_4^+(aq) + 3H_2O(\ell)$

3.243 $Hg + 2NO_3^- + 4H^+ \rightarrow Hg^{2+} + 2NO_2 + 2H_2O$

NOTES

NOTES

NOTES

NOTES

NOTES

NOTES

NOTES

NOTES